Tangled Up

Michael Hornberger

First published by Canbury Press 2025

This edition published 2025

Publisher: Canbury Press (www.canburypress.com)

14 Beresford Rd, London, KT2 6LR, United Kingdom

EU Authorised Representative: Easy Access System Europe

- Mustamäe tee 50, 10621 Tallinn, Estonia, gpsr.requests@easproject.com

Printed and bound in Czechia by Finidr

Typeset in Athelas (heading), Futura PT (body)

All rights reserved © Michael Hornberger

Michael Hornberger has asserted his right to be identified as the author of this work in accordance with Section 77 of the Copyright, Designs and Patents Act 1988

This is a work of non-fiction. It contains general health information, not specific medical advice. Neither the author nor publisher accept any liability for any action arising from its reading. If you have a concern about your health, please consult a medical professional.

FSC® helps take care of forests for future generations.

Hardback ISBN 9781914487422

Ebook ISBN 9781914487415

Tangled Up

The History and Science of Alzheimer's Disease

Michael Hornberger

Canbury Press

CONTENTS

Prologue	7
PART 1. Alois & Auguste (And A Bit Of Oskar)	**15**
1. Dr Hoffmann's Vision	17
2. Mrs Deter	23
3. Alois Alzheimer	33
4. Munich Via Heidelberg	39
5. 1906	41
6. The Tübingen Aftermath	45
7. Fischer's Disease	49
8. Alzheimer's Legacy	53
Part 1. Summary	55
PART 2. Memory And Space	**57**
9. 'You Are Not My Type'	59
10. The Art Of Memory	63
11. 'Secret' Memory Techniques	67
12. 'Attention!'	73
13. 'Enter The Seahorse'	77
14. Encoding, Storage, Retrieval	81
15. 'It's All About Access'	85
16. The Imaginary Mrs A	89
17. Spatial Disorientation	97
18. 'Lost In Space'	101
Part 2. Summary	107
PART 3. Amyloid And Tau	**109**
19. Glorious Proteins	111
20. Amyloid	113

21. Beta-Amyloid Formation — 115
22. 'Location, Location, Location... And Timing' — 119
23. Tau — 121
24. 'Who Ordered All The Phosphate?' — 125
25. Tau 'Infection' — 129
26. The 'Emergence' Of The Disease — 133
27. Biomarker Technology — 137
28. Amyloid Treatment Approaches — 143
29. Tau Treatment Approaches — 151
Part 3. Summary — 155

PART 4. Genetics And Lifestyle — 157
30. Genetics 101 — 161
31. Familial Alzheimer's Disease — 165
32. 'Modifiable' Risk Genes — 173
33. A Word On Genetic Testing — 179
34. Non-Modifiable Lifestyle Factors — 181
35. Modifiable Lifestyle Factors — 187
Part 4. Summary — 205

PART 5. Rarer Forms Of Alzheimer's Disease — 209
36. Frontal Variant Alzheimer's Disease — 211
37. Posterior Cortical Atrophy — 217
38. Logopenic Variant Primary Progressive Aphasia — 223
39. Corticobasal Syndrome — 229
40. A Final Word On The Rarer Forms Of Alzheimer's — 235
Part 5. Summary — 237

Acknowledgments — 239

PROLOGUE

This is not a book about dementia. This is a book about Alzheimer's disease.

What's the difference, you might ask?
The difference is that dementia is an umbrella term for all types of dementia, with Alzheimer's disease being the most common form.

Why would one need a book on Alzheimer's disease and not a book on dementia?
There are already many excellent books on dementia available. However, to my knowledge there is none on the science behind Alzheimer's disease specifically.

Why does it matter?
If we want to understand the science behind dementia, we need to look at each type of dementia specifically, as the science for each type of dementia is quite different. We need to understand how a particular type of dementia develops in the brain, what changes it causes in the brain and how this results in the symptoms we see for this type of dementia. In essence, what is the science behind each type of dementia?

What is this book about?
This book focuses on the science (and history) of Alzheimer's disease. Specifically, we will explore in detail how the brain changes in Alzheimer's disease cause the symptoms of the disease and how new, upcoming treatments will deal with these brain changes. We will also explore the risk factors for Alzheimer's disease and how we can potentially reduce them. We will start by travelling back in history to find out how Alois Alzheimer and Oskar Fischer 'discovered' the disease.

Who is this book for?
For everyone, and I mean everyone, who is interested in the science behind Alzheimer's disease. No prior scientific knowledge should be required to understand the book, as I have gone to great lengths to explain the scientific concepts behind each important aspect. So, if you are an interested lay person, a person with dementia, a paid or unpaid carer, a family member, a dementia advocate, a healthcare professional or even a clinician, this book is for you. However, although this book will cover the basic science behind Alzheimer's disease, it will not go into the science of dementia care, for which there are already some excellent books available by lay and professional carers. But if you are curious about how the changes in the brain cause the disease and its consequent symptoms and how we can potentially prevent them, then this book is for you.

Why this book now?
Communication of science has been a passion of mine for nearly two decades. The reason I became interested was that I noticed how little scientific information was out there to provide people with detailed, but understandable scientific information on Alzheimer's disease. What I found instead was a choice between either very generic scientific information, of the sort provided by websites, such as 'it causes changes in the brain affecting your memory'; or scientific publications filled to the brim with jargon and acronyms, making it impossible for lay people to understand what these articles were discussing.

I think there is a compromise between these two extremes, which allows a general reader to get a more detailed understanding of the science behind Alzheimer's disease. Of course, I am not the first to think of that. Most dementia charities and organisations already provide such lay-friendly dementia science information. However, I could not find a book which compiled all the information in one place. So, I decided to write my own, based on cutting-edge scientific evidence at the time of writing. 'At the time of writing' is an important qualifier, as scientific progress is continuous, and some findings might change soon.

The other caveat I should add that this book is not for the reader who wants to get a brief overview of dementia. To those readers, I recommend consulting the dementia charity websites in their country or Alzheimer's Disease International (see the links at the end of the book). This book is written for people who want to get a detailed understanding of the science behind Alzheimer's disease. If you stick with it, you will be rewarded by having an in-depth understanding of how Alzheimer's disease develops in the brain, how it causes the typical symptoms, how we can treat the disease and finally how we can even potentially prevent it. You will also understand this book's title.

In essence, you will understand the fundamentals of the disease which will help you understand any scientific findings you might come across in the future.

I have structured the book in five parts to guide you through the science of Alzheimer's disease: I) Alois and Auguste (and a bit of Oskar); II) Memory and Space; III) Amyloid and Tau; IV) Genetics and lifestyle; V) Rarer forms of Alzheimer's disease.

Let's go through each of them briefly so you know what to expect.

1. Alois and Auguste (and a bit of Oskar) – The first part of the book will delve into the history of Alzheimer's disease. It will give us an understanding of how the disease was 'discovered'. We meet Dr Alois Alzheimer when he first encounters Mrs Auguste Deter – the 'first' person diagnosed with Alzheimer's disease. Importantly, we can also follow

Alzheimer's discovery of how specific protein changes in Mrs Deter's brain caused her symptoms. We all have heard of a disease called Alzheimer's disease but most of us do not know who Alzheimer was. It seems very fitting that we start a book on Alzheimer's disease by meeting Alois Alzheimer first. The other person we will come across is Dr Oskar Fischer, who discovered the 'Alzheimer' protein changes at the same time as Alzheimer did. Unfortunately, Fischer has been forgotten by history, despite his important contribution and it is time to rectify this omission.

2. Memory and Space – The second part of the book will explore how memory and specifically episodic memory works. The reason why there is a whole section dealing solely with memory is that, while there are also other cognitive changes, episodic memory symptoms are a key feature of Alzheimer's disease. Later in the book, we will cover some of the other cognitive symptoms found in the rarer forms of the disease. To understand the typical memory symptoms in Alzheimer's disease, such as forgetting events or 'living in the past', we first need to understand how memory actually works in the brain. This will make it clear why only certain memories are affected by Alzheimer's disease and why we should only worry about some memory changes but not others. The final part of the book will explore the other common but under-recognised early symptom of Alzheimer's disease – spatial disorientation. Spatial disorientation is another key feature of Alzheimer's disease and explains why people with the disease are at a greater risk of getting disorientated or even lost. Again, we will explore how spatial navigation works in the brain and how the disease affects these processes.

3. Amyloid and Tau – In Part 3, we are now ready to take a deep-dive into the biology of the disease – the so-called

pathophysiology. This part of the book is called Amyloid and Tau, already giving away that these two proteins are key to the development of Alzheimer's disease. Admittedly, this is a large and complex part of the book and I would recommend reading the chapters bit by bit to digest all the information given, as there is a lot of it. First, we are going to explore the amyloid protein and what it usually does, before delving into how it goes awry in Alzheimer's disease. Next up, we will do the same for the tau protein by exploring its normal function and what happens in Alzheimer's disease. We will then briefly explore how we can measure amyloid and tau in the brain and the possible future diagnosis of Alzheimer's disease. The final chapters will explore the different treatment approaches for amyloid and tau which are currently taking place in clinical trials, before briefly exploring why most clinical trials of Alzheimer's disease have failed so far – with a shimmer of hope recently emerging.

4. The fourth part of the book will look at another complex and large topic – genetics and lifestyle factors which determine our risk for Alzheimer's disease. First off, we will go through the genetics, which is often a big worry for people who have had relatives affected by the disease. Have I inherited the genes from my relatives that might lead to me also developing Alzheimer's disease in the future? The short and reassuring answer is that it is unlikely that our genes alone will cause us to develop Alzheimer's disease. However, we will explore how our genes can influence our risk for Alzheimer's disease. Still, genetics is its own vast universe in science and to explain how genetics affect Alzheimer's disease, I have added a chapter (Chapter 30 – Genetics 101), which explains some basic concepts of the science of genetics. The latter half of this part of the book will explore how our lifestyle choices as well as our environmental factors affect our risk for Alzheimer's disease.

We will then explore how some of these lifestyle risk factors can actually be modulated, thereby reducing our future risk.

5. By Part 5 of the book, you hopefully know a lot about the science behind the most common form of Alzheimer's disease, causing the typical memory and disorientation problems. However, there are several rarer forms of Alzheimer's disease which cause a variety of different symptoms from vision to language to motor symptoms. Unfortunately, these rarer forms of Alzheimer's disease are less recognised and understood. The last part of the book will therefore go through each of these rarer forms of Alzheimer's disease – the exception being familiar Alzheimer's disease, which we will have covered in the genetics section of the book. Still, understanding these rarer forms is important in understanding how Alzheimer's disease can present in quite different shades of symptoms and that it is not only about memory and orientation in Alzheimer's disease. If you have never heard of any of those rarer forms of Alzheimer's disease, I hope this part of the book will broaden your horizon to know that other forms of Alzheimer's disease exist and what they might look like.

Each part of the book can be read separately. However, for a more holistic insight into the science behind Alzheimer's disease, I recommend reading all parts in the order provided. In particular, it is worth understanding how Alzheimer 'discovered' the disease and how memory symptoms occur, before then going into the protein and gene details, which are the most complex parts of the book.

Finally, I want to invite my lay readers to provide feedback on anything in the book. I have striven to make the science behind the pathophysiological processes that cause Alzheimer's disease intelligible for general readers, however, there is always room for improvement. I also hope my professional colleagues will forgive me for simplifying some of the pathophysiological processes. In reality, of course, the science is far more complex. Nonetheless, I have tried hard not to 'over-simplify', but rather to convey the essence of the science behind Alzheimer's disease. But clearly this is debatable, and I would be keen to hear from you if you have some suggestions on how to convey some of the pathophysiological processes in a better or more accessible way.

But now, let's get started with the history and science of Alzheimer's disease.

PART 1.
ALOIS & AUGUSTE
(AND A BIT OF OSKAR)

When Dr Alois Alzheimer walked into the 'Hospital for the Mentally Ill and Epileptics' near Frankfurt in Germany, on the morning of 26 November 1901, he was quite unprepared for the events which would unfold from this fateful day. This was a day that would not only have a profound effect on the next five years of Alzheimer's life but would go on to impact the lives of millions of people over the coming decades.

Alzheimer had joined the hospital in 1888 as a junior doctor [German: 'Assistenzarzt']. In the prior 13 years, he had enjoyed his work and was well-liked by his colleagues and the head of the hospital Prof Emil Sioli. It was Sioli and his vision for the new hospital which had persuaded Alzheimer to work there. Not only was it a state-of-the-art facility but its leadership was committed to new methods in the care and treatment of mental health conditions, including – importantly – gaining a better scientific understanding of mental health changes.

This drive towards developing better care for people with mental health problems and understanding scientifically how these diseases started suited Alzheimer perfectly. He had realised early on that excellent medical care goes hand-in-hand with a deep understanding of the underlying causes of the problems his patients faced. The new hospital and Sioli's leadership perfectly aligned with this vision of a new approach towards mental health.

1.
DR HOFFMANN'S VISION

For centuries people who had mental health conditions were often locked away in institutions, known colloquially as 'madhouses'. This had also been the case in Frankfurt, where such a 'madhouse' had existed from the 1600s until the early 19th Century. However, these so-called institutions were merely prisons which locked away people with mental health conditions from the general population. The prison-like character of such places was further compounded by the fact that they consisted of filthy, locked cells holding many people, some of them chained to beds or walls. These institutions were usually manned solely by an overseer and a guard. A doctor had a remote directorship position, visiting only on a 'need-to' basis.

If you ever want to get an insight into how such institutions looked and felt, then I would recommend a visit to – or look up on the internet – the Narrenturm (Tower of Fools) in Vienna. The Narrenturm is the oldest surviving 'psychiatric hospital' in the world. However, there is not much of a hospital there. The structure of the building tells us exactly what its intended purpose was – not to treat people with mental health problems but to incarcerate them. Today the Narrenturm is a museum for the clinicopathological collection of the University of Vienna, but a visit to the building and an immersion in its atmosphere provides us with a gruesome insight into how horrific conditions were for people with mental health problems. It was no different in Frankfurt, 445 miles away.

The inhumane conditions of the institution for the people with mental health conditions in Frankfurt shocked one doctor, Dr Heinrich Hoffmann, who decided to change things for the better when he became the director of the Frankfurt 'madhouse' in 1851. Since Hoffmann was not only a well-connected doctor but also the author of a highly successful children's book (Struwwelpeter, which is still in print in Germany today), he had excellent connections within the higher echelons of Frankfurt society. Hoffmann, therefore, decided not to wait for the city of Frankfurt to provide the funds for a new, 'state-of-the-art' institution for the people with mental health conditions. Instead, he organised a fundraising campaign for the new building, drawing on the more affluent classes of Frankfurt society. The campaign was a resounding success and raised enough money for the doctor to build a brand new mental health institution in Frankfurt.

Hoffmann didn't do things half-heartedly. So, instead of immediately starting to build his new hospital with the new funds, he first toured Germany and selected European countries to examine how other state-of-the-art mental institutions were laid out and run. He returned to Frankfurt several months later, having seen many such newly built institutions and his mind was made up about what the town needed to replace its 'madhouse'. Armed with this information, Hoffmann convinced the city council of Frankfurt to build a new purpose-built institution for the people with mental health conditions and those with epilepsy. The city acceded to his request reluctantly, as it had 'only' planned to replace the existing institution. But Hoffmann's prominence in Frankfurt society and the substantial funding brought much public scrutiny to bear on the council's decision. Instead of replacing the old 'madhouse' in the middle of the city, the new institution for the mentally ill was built on a hill (Affensteiner Hügel) just outside of Frankfurt, surrounded by large parklands. The original building no longer exists and the space is today occupied by buildings of the University of Frankfurt.

To some, the decision to locate the new hospital just outside Frankfurt's walls was a deliberate move by the city council to place the people with mental health conditions as far as possible from city life. For Hoffmann it did not matter – in fact, he thought the location of the new building within parklands was ideal. He had learned during his travels that patients recover better by having access to nature since it 'calmed their nerves'. Therefore, having a building where patients could – under supervision – roam gardens or even just have views of nature would help improve their care and treatment. The consent for the new building in its suggested location was therefore given and building work started shortly thereafter. The new building was completed in 1864 and Dr Hoffmann was appointed its first director. But completing the building was not enough for the doctor.

He had not lost sight of his original vision that a state-of-the-art building was only the first step in better care of the patients. Equally important for him was improving the hospital's treatment and care for the people with mental health conditions. But this was not all. He also wanted to link the care and treatment of mental health conditions to science, which his travels had shown could improve care significantly. It also chimed with Hoffmann's own beliefs that mental or physical changes in the brain could explain most mental health conditions. This was a truly revolutionary notion at a time when mental health problems were seen by the majority of the population as being something akin to simply difficult behaviour or, worse, a form of possession by inner demons.

To embed scientific discipline within the hospital, Hoffmann allocated some rooms in the building for research early on in the planning stages. He did not specify the usage of those rooms but it was clear that he was keen to create a space for cutting-edge mental health science, including the fledgling fields of microscopy and histology (the microscopic anatomy of cells and organ tissues). The rooms on the first floor would allow doctors to investigate the scientific causes of the mental and physical disorders suffered by admitted patients. While this was an approach that had already been trialled at some of Europe's leading medical centres, it was truly revolutionary for

Frankfurt, which at the time did not have a medical school or even a university.

Hoffmann's other passion was for the improved care and treatment of mental health problems. Instead of simply locking people away or even chaining them to walls or beds, he became a strong supporter of the new 'non-restraint' principle in the treatment of mental illness. The 'non-restraint' principle states that the restraining or chaining of people with mental health problems should be avoided as much as possible, as long as the safety of the person is guaranteed. It is now commonly used in the treatment of such problems, but at that time was almost unprecedented.

Hoffmann came across the new non-restraint method when he visited England to carry out his background research. Here, selected institutions had been using this technique under the influence of William Tuke's pioneering work in the 18th Century. Hoffmann must have been deeply impressed by this approach, as he designed the whole building with the non-restraint principle in mind. This meant that instead of small, locked cells there were large bedrooms in which patients could walk freely. There were also separate sections for people with different mental health symptoms. For example, there was a special section for manic patients, which also had an outside courtyard, allowing them to pace around as much as they wanted. Similarly, patients with anxiety had a garden and were allowed to roam the parkland surrounding the hospital under supervision.

Finally, informing his passion for science and non-restraint, was Dr Hoffmann's humanist outlook, which held that all people and patients were equal, regardless of their economic, cultural or religious background. He therefore made the new mental health institution accessible to Jewish patients for the first time, something which, until then, the Frankfurt 'madhouse' and many others had refused to do. Hoffmann also consciously included the wealthy Jewish families of Frankfurt in his fundraising campaign for the new hospital, meaning that the Jewish community in Frankfurt had financially contributed to the new building, leaving the city council little choice but to admit Jewish patients as the Jewish community now part-owned the building. It's clear then that the contribution Hoffmann made to the discovery of Alzheimer's disease should not be underestimated as he created the ideal environment for Alzheimer's ground-breaking work.

For the next 22 years Hoffmann ran the new institution as a director, overseeing great improvements in the care and treatment of people with mental health problems in Frankfurt. When he stepped down as director in 1888, he was keen to have someone carry on his work and, more importantly, expand it further. He found the right person in Dr Emil Sioli, a physician who fully supported Hoffmann's vision. Not only was Sioli a follower of Hoffmann's methods, but he was also part of the first generation of psychiatrists (or 'nerve doctors' as they were called then), which emerged at that time in Germany. While Sioli's specialism in the relatively new field of psychiatry made him the perfect choice for the new director, he also wanted to build up a more substantial clinical team which could help him to consolidate the institution's earlier work.

So, shortly after he took up his new role, Sioli appointed a young 24-year-old junior doctor – Alois Alzheimer to assist him in his work at the hospital. To have a more experienced doctor to support Alzheimer, Sioli also appointed Dr Franz Nissl as a consultant. Both Alzheimer and Nissl not only pursued the same ethos as Hoffmann and Sioli but also pushed the boundaries of science in mental health. Little were they aware that their future work would have such a long-lasting impact and ultimately make medical history. Another person who would share an unwitting role in their revolutionary work was a middle-aged woman named Auguste Deter.

2.
MRS DETER

On 26 November 1901, Alzheimer was reviewing a list of cases that had been admitted to the hospital the previous day. The case of a 51-year-old woman caught his eye. The referral letter from her family doctor Dr Leopold L (unfortunately we only have the first letter of his surname) mentioned that Auguste Deter had been 'suffering for some time from memory problems, delusions, sleep problems, restlessness' based on the information given by her husband August Wilhelm Karl Deter. Dr L diagnosed her with 'chronic brain paralysis' (German: 'chronische Hirnparalyse'), a medical term no longer in use. Literally translated, it means that her brain seemed to have stopped working normally for a prolonged period of time. Dr L suggested that she should be treated at the psychiatric hospital. The admitting doctor, Dr Nitsche, also noted that during Mrs Deter's admission, her husband had reported that they had been happily married since 1873 and had one daughter. His wife had always been very active and busy, although at times rather nervous. She had been in general good health and Dr Nitsche also noted that there was no history of drinking alcohol or any venereal diseases.

Although it might seem strange to our modern ears for a psychiatrist to ask for a history of sexually transmitted diseases (STDs), this was routine at the time. STDs such as syphilis can eventually cause severe changes to a sufferer's mental health and, prior to the invention of penicillin in 1910, around 10% of the population was estimated to have syphilis. Unsurprisingly, research into neurosyphilis was a considerable part of the fledgling field of psychiatry, so it made sense for Nitsche to enquire about this. However, there was no indication that Mrs Deter had syphilis or any other sexually transmitted disease.

After consulting Nitsche's admission notes, Alzheimer examined Mrs Deter himself. After a brief introduction, he asked her a series of questions, some of which are still used in the patient examinations these days, such as where she thought she was, what her name was and where she lived. Due to Alzheimer's diligent note-keeping, we can actually follow their verbatim conversation:

Alzheimer: 'What is your name?'
Deter: 'Auguste.'
A: 'Family name?'
D: 'Auguste.'
A: 'What is the name of your husband?'
D: 'I think Auguste.' [It is unclear whether she was really referring to her husband who, confusingly, was called August Wilhelm Karl]
A: 'Your husband?'
D: 'I see, my husband...'
A: 'Are you married?'
D: 'To Auguste.'
A: 'To Mrs Deter?'
D: 'Yes, to Mrs Deter.'
A: 'How long have you been here?'
D: 'Three weeks.'
A: 'What do I have in my hand?'
D: 'A cigar.'
A: 'Correct. And what is that?'
D: 'A pencil.'
A: 'Thank you. And that?'
D: 'A fountain pen.'
A: 'Also correct. What is this, Mrs Deter?'
D: 'Your wallet, doctor.'

A: 'Yes. Correct. And that?'
D: 'A book.'
A: 'And what lies beside my notebook?'
D: 'A keyring.'
A: 'What does it consist of?'
D: 'Many single keys.'

From this verbatim exchange, we can see that Alzheimer is probing different aspects of Mrs Deter's memory. In particular, he probes for her personal events memory (so-called episodic memory) by asking her name, her marital status, duration of hospital stay, but he also asks her more everyday questions (concerning so-called semantic memory) by asking for the names of objects. From these first few exchanges, it is clear that Mrs Deter is quite confused and disorientated, as she had only been admitted the day before, but believes she has been in the hospital for three weeks. Contrast this with her knowledge of the objects Alzheimer is showing her. She clearly knows these items, even the cigar, although it seems odd to us now that a doctor is smoking a cigar while examining a patient.

To conclude his first examination, Alzheimer's asks her to write several things, including her name and address. One symptom he instantly notices is that in the middle of writing her name, she has forgotten what she was meant to write. This is a memory symptom, which piques Alzheimer's interest. He notes it down as 'amnestic writing disturbance' (original German: 'amnestische Schriftstörung') in his report, while also noting that she still understood the meaning of the words and could name everyday objects without any problems.

Over the next three days, Alzheimer seems to have thought more about Mrs Deter's memory symptoms. He, therefore, decides to conduct a more thorough interview with his patient.

Alzheimer: 'How are you?'
Deter: 'It is one like the other. Who brought me here?'
A: 'Where are you?'

D: 'At the moment; I have said it before, I do not have any means. One just has to – I do not know myself – I do not know – oh dear, what is this then?'
A: 'What is your name?'
D: 'Mrs Deter, Auguste'
A: 'When were you born?'
D: 'Eighteen hundred and...'
A: 'What year were you born?'
D: 'This year, no, last year.'
A: 'When were you born?'
D: 'Eighteen hundred – I do not know...'

Another common symptom in Alzheimer's disease is how disorientated people can be regarding time and place. As we can see in the previous exchange, Mrs Deter can't remember the year she was born and also gets her birthday confused with the current and previous year.

A: 'What did I ask you?'
D: 'Well, Deter Auguste...'
A: 'Do you have children?'
D: 'Yes, one daughter.'
A: 'What is her name?'
D: 'Thekla!'
A: 'How old is she?'
D: 'She is married in Berlin, Mrs Wilke.'
A: 'Where does she live?'
D: 'We live in Kassel.'
A: 'Where does your daughter live?'
D: 'Waldemarstreet – no, different...'
A: 'What is the name of your husband?'
D: 'I don't know...'
A: 'What name does your husband have?'
D: 'My husband isn't here at the moment.'
A: 'What is your husband called?'
D: [suddenly quickly]: 'August Wilhelm Karl – I do not know if I can tell you like that.'

A: 'What is your husband['s profession]?'
D: 'Administrator [original German: 'Kanzlist'] – I am so confused, so confused – I cannot.'

In a prior part of their conversation, Alzheimer had probed for information about the people close to Mrs Deter, asking for her husband's and daughter's names and details. From the conversation, it becomes clear that she can remember her daughter and her name. She also knows that the daughter is now married under a different surname and lives in Berlin. However, she gets this information confused when asked for her daughter's address, in answer to which she actually names the street where she grew up in Kassel (a town 120 miles/200km north of Frankfurt). She has difficulty remembering her husband's name, but once the memory is triggered, she quickly remembers his full name and profession. This suggests that her memories are intact but she cannot access them anymore, which is a very common symptom in early Alzheimer's disease. Therefore, people are often unable to remember information but once the memory is triggered for some reason, they can recall the required information.

Let's rejoin them:
A: 'How long have you been here?'
D: 'About two days...'
A: 'Where are you?'
D: 'This must be Wilhelmshöhe...'
A: 'Where is your flat?'
D: 'Well, Frankfurt am Main...'
A: 'In which street?'
D: 'Waldemarstrasse... Not, it has to be another, just wait a moment – I am just so, so...'

Again, she seems disoriented as to time and place but more importantly, we can see here the typical symptoms of a person with Alzheimer's disease 'living in the past'. Mrs Deter seems to think that she is back in Kassel, where she was born and grew up, although she actually lived most of her life in Frankfurt. Instead of realising that she is in Frankfurt, she thinks that she is on the Wilhelmshöhe, which is a hill close to Kassel, similar to the Affensteiner Hügel, where she really is. She must have deduced from the view she had from the institution that she was on top of a hill and because she believes that she lives in Kassel, 'This must be Wilhelmshöhe...'

A: 'Are you sick?'
D: 'Just down there, along the spine...'
A: 'Do you know me?'
D: 'I think you have seen me twice already. No, my apologies, I just cannot...'
A: 'What year is it?'
D: 'Eighteen hundred...'

In this section of the interview, Alzheimer probes into whether she has any insight into her own symptoms, which she doesn't seem to have as she mentions problems with her back instead of her mental state. Asking her if she remembers him is also a good way for him to gauge how much she remembers from their previous meetings and she actually correctly indicates that they have seen each other twice before but then hesitates and is not sure any more. It gives Alzheimer an indication that she remembers some recent aspects of her life, but even if she does, she is not sure of herself or confuses them with other events. Finally, he asks her again for the year. Now, this is not because Alzheimer has actually forgotten that he has asked her this already or that he is simply cruel. Instead, he wants to check if she remembers the information now, after a brief delay.

We rejoin them for the final part of this interview:
A: 'What month is it?'
D: 'The second month.'
A: 'What are the months called?'
D: 'January, February, March, April, May, June, July, August, September, October, November, December.'
A: 'Which month is it now?'
D: 'The eleventh.'

A: 'What is the eleventh month called?'
D: 'The last one then – no, not the last one...'
A: 'Which one?'
D: 'I do not know...'
A: 'What colour is snow?'
D: 'White.'
A: 'Coal dust?'
D: 'Black.'
A: 'The sky?'
D: 'Blue.'
A: 'A meadow?'
D: 'Green.'
A: 'How many fingers do you have?'
D: 'Five.'
A: 'Eyes?'
D: 'Two.'
A: 'Legs?'
D: 'Two.'
A: 'How many pennies [original German: Pfennige] are there in one pound [original German: Mark]?'
D: 'One hundred.'

...

Here again, as in their first interview, Mrs Deter's semantic memory is excellent. She can clearly remember facts and knowledge of the world, while her episodic memory shows significant changes. Alzheimer asks her once again to write her name but after writing 'Mrs' she has forgotten what she was meant to write. Similarly, he asks her to read a brief passage and she struggles to read from one line to the next and often has trouble remembering the previous line. As a final part of the examination, Alzheimer conducts a physical examination and reports that her pulse is normal but weak and there are no heart murmurs. For the neurological examination, he reports that 'both eye pupils reacted normally to light. Sticking out the tongue, it appeared very dry, likely due to some slight dehydration.' He also notes that her spoken language appears normal, even though she is hesitant and that she wears dentures. All in all, her physical and neurological examination appears fine but she clearly has significant cognitive symptoms, especially relating to episodic memory.

Over the next two weeks, Alzheimer conducts several interviews with Mrs Deter, often asking the same questions as before. Where are you? What year is it? Where do you live? Some days her answers are more lucid than on others when she is very taciturn or even gets distressed by his questions. Outside of the interviews the nurses have noticed that she is also very nervous, and sometimes becomes highly agitated. She is therefore sometimes placed within a single-occupancy room, instead of the larger bedrooms, which are shared between 20-30 people. The single-occupancy room seems to calm her down. Based on his assessments and her agitation, Alzheimer confirms that Mrs Deter should remain at the hospital under observation. It's also likely that he prescribed hydrotherapy, which was the main treatment method at the hospital during that time. Hydrotherapy, which originated in England, required patients to take prolonged baths in warm water (the water was consistently kept at 34°C). The theory behind it was based on the findings that bathing could be calming for people with mental health problems and that it might therefore also change their symptoms. At that time, modern psychiatric hospitals such as the one Alzheimer worked at, didn't just have halls with beds but also large bathing facilities for the patients, which were separate for men and women.

One curious thing worth noticing is that Alzheimer painstakingly makes notes of all the interviews with Auguste Deter and also asks a photographer to take some photos of her at the beginning of 1902 – the most famous one showing her sitting on the bed in her white gown, which was the prescribed clothing for all patients on the wards. Alzheimer clearly paid close attention to Mrs Deter, but it is not clear whether Alzheimer instantly knew whether Mrs Deter was a 'special' patient. On the one hand, it is unlikely, as Alzheimer and other clinicians were seeing dozens of patients every week, each with their own symptoms and no diagnostic criteria existed at that time for most psychiatric diseases. It must have been very difficult to identify whether one particular case was a variation of the other cases or a completely different disease.

This can still be a problem today in difficult cases. However, clinicians these days have clear diagnostic criteria for most diseases and there is a wealth of in-depth knowledge of the diseases. In contrast, for Alzheimer and his contemporaries, the variety of symptoms they saw must have been bewildering. Most people would have only been admitted to an institution when they were quite advanced in the progression of their diseases and the family could no longer cope with them. This was also the case with Mrs Deter. Some readers may have already spotted that Mrs Deter was already in the moderate, if not advanced, stages of Alzheimer's disease. Seeing patients, such as Mrs Deter, in the later stages of the disease made it even harder for Alzheimer and his contemporaries to diagnose them. The reason for this is that for progressive neurodegenerative diseases, like Alzheimer's disease, the symptoms of the disease are clearest at the start when only specific brain areas are affected. However, once the disease spreads, there is often a larger range of symptoms, which overlap with other diseases. Therefore, seeing people in the more advanced stages of the disease without any diagnostic criteria was very challenging if clinicians wanted to diagnose those different diseases and conduct research into them.

Still, Alzheimer saw something in Mrs Deter, as he paid close attention to her, indicated by his extensive note-taking and the photographs. He was particularly taken by her 'writing amnesia', that she forgot mid-sentence what she was writing, while at the same time being very lucid in her knowledge of objects and facts. In most other psychiatric cases, it's more likely that they would be confused on a general level and would have had problems answering any of Alzheimer's questions. Nonetheless, Mrs Deter, despite being quite confused, showed this dissociation of episodic and semantic memory, which piqued his interest. Another reason for his particular attention to her case might be that he had seen a similar case three years prior, in 1898. He even published details of that case, along with other cases of psychiatric diseases, but we have very little information on the actual person and the detailed symptoms in that original case. Still, that previous encounter might have piqued Alzheimer's interest when he met Mrs Deter. He likely made the connections between the patient he had seen in 1898 and Mrs Deter, and thought that these were not isolated cases but potentially a different disease with the same underlying symptoms. There seemed to be a pattern to the specific episodic memory symptoms in both cases, which was worth investigating further. This could explain Alzheimer's painstaking note-taking in the case of Mrs Deter. Instead of seeing her as the first person with Alzheimer's disease, we should regard Mrs Deter as rather the first well-documented case of Alzheimer's disease. Alzheimer and other clinicians had probably seen and even reported similar cases before this case. But what Alzheimer did next was critical in making Mrs Deter the first person to be diagnosed with Alzheimer's disease.

3.
ALOIS ALZHEIMER

Alois Alzheimer was born on 14 June 1864 in Marktbreit, a small town in the north of Bavaria, in modern-day Germany, about halfway between Frankfurt and Nuremberg. He was one of three children from the second marriage of his father Eduard, a notary, and his mother Barbara, a housewife. Alzheimer went to the local primary school in Marktbreit before the family moved to Aschaffenburg – a larger town 100km away – where the children were able to attend a better secondary school. During this time, Alzheimer seemed to have shown an avid interest in the natural sciences and it was decided that he should study medicine. He began studying medicine in Berlin and Tübingen, but conducted the majority of his studies at the nearby Medical School in Würzburg.

There is no evidence that Alzheimer was a particularly gifted student at university. On the contrary, he was more noted for his mischievous behaviour, which led to several reprimands and even talk about him being excluded. This same mischievous streak seemed to have continued all his life and he was known to be fond of practical jokes, not only on his children but even on his colleagues, and had an eccentric streak; he kept a pet monkey at home.

His playful nature might explain why Alzheimer was a member of a student fraternity (schlagende Verbindung; Corps Franconia), of the sort that were very popular at the time in Germany. These student fraternities had simultaneously liberal and nationalistic tendencies, and played a pivotal role in the eventual formation of Germany in 1846 – which had previously been a conglomeration of kingdoms (Prussia, Bavaria, Saxony) and principalities. Besides their political element, the fraternities were also highly social groups and known for duelling and practical jokes. The duelling consisted of sabre fighting without any head protection, with most members proudly showing the scar(s) from these confrontations to demonstrate their bravery and prowess. Alzheimer himself had a prominent scar from such a duel, on the right side of his face, which can be seen on some of his portraits. Despite leading to the occasional reprimand from the university authorities, membership of these fraternities also provided Alzheimer with friendships throughout his life. So, while it would also be fair to describe Alzheimer as a 'bon vivant', he was also known to work extremely hard, being meticulous in everything he put his mind to. Or to put it in modern parlance: he worked hard and played hard.

In his final year of medical school in 1887, Alzheimer had to conduct a scientific dissertation, which would have a profound impact on his future career and the discovery of the disease that would bear his name. During the dissertation, he worked at the anatomical institute of the University of Würzburg as part of a group taught by Professor Albert von Koelliker. Koelliker was a leading expert in anatomy, physiology and histology at the university, and was in particular interested in a new cutting-edge field of science – microscopy.

Microscopy allowed scientists, for the first time, to investigate cell structures and human tissue (histology) in detail. Microscopy not only allowed for the study of cells and human tissue in healthy people but also how they might change in disease. This led to the new field of pathology, which investigated how diseases affect human cells and tissues via microscopic techniques. Prior to that, clinicians and anatomists relied only on what they could see with their eyes (macroscopic) to analyse the changes that diseases made to the body. The arrival of the microscope allowed us to complement these macroscopic observations with microscopic observations. Not only can we now observe changes to patients by looking at them, we can also investigate how their cells or tissue are affected by the disease when looking through the microscope. It was a revolutionary advance in medical sciences, and one which fascinated Alzheimer.

Koelliker himself never became well known for his research but his teaching on microscopy influenced a whole new generation of students. Many of his students used this cutting-edge technology to conduct future groundbreaking research, including Alzheimer. Alzheimer's dissertation itself was on the histology of the ceruminous glands, which are the glands in our ears that produce earwax. While the topic itself was not relevant for Alzheimer's future career and discoveries, he seemed to have been taken, if not hooked, by microscopy, as he spent countless hours over the microscope. This hard work and meticulousness were awarded with a distinction for his dissertation. The microscope not only provided a successful end to Alzheimer's medical studies but it shaped his whole future career as a clinician.

Microscopy was also one of the reasons why Alzheimer joined the new hospital in Frankfurt, as the city was one of the leading microscopy centres in Germany at that time. At the nearby Senckenbergischen Pathological Institute, Carl Weigert and Ludwig Erdinger were pioneers of the new method of investigating tissue and identifying diseases via microscopy. After arriving in Frankfurt, Alzheimer often visited their institute and learned advanced microscopy techniques, such as how to best prepare tissue for microscopy. He also learned how to stain (a form of dyeing) the tissue for identifying different structures under the microscope.[1] Weigert and Erdinger pioneered the use of different stains for the microscopy of tissue, which made different structures

---

[1] Staining the tissue is a critical step for microscopy – even today. Most tissue, if it hasn't been stained, will look extremely underwhelming when put under the microscope. I remember this myself, having received a microscopy set one Christmas as a child. Excitedly, I put onion skins under the microscope – snatched from my mother's kitchen during dinner preparation. But the view of the skins was extremely disappointing. Only later, when I read the instructions that came with the set, did I realise that several dyes were provided. These dyes or stains highlighted the tissue structures in vivid colours, making them distinguishable to the viewer.

in the tissue visible and distinguishable. This made it possible to compare tissues from healthy people with tissue from patients, in order to identify which changes inside or outside cells were caused by the disease. This was a truly revolutionary approach which is now used everywhere every day to diagnose patients with all kinds of diseases. The innovative environment was an ideal one for Alzheimer to pursue his passion for microscopy while working as a clinician. He converted one of the scientific rooms at the hospital, which Hoffmann originally set up, into a microscopy suite. The new microscopy suite would allow for the study of the tissue changes of the patients seen at the hospital. Despite his enthusiasm for microscopy, Alzheimer was at that time no expert in the subject. But crucial microscopy expertise soon arrived at the hospital in the shape of Dr Franz Nissl.

Like Alzheimer, Nissl was a psychiatrist, but his real passion was in the field of microscopic histopathology – the identification of diseases in the tissue under the microscope. It is unclear what initially attracted Nissl to the hospital in Frankfurt, as he came from one of the leading mental illness institutions in Germany at that time, Munich. In Munich, Nissl had learned microscopy under the tutelage of Professor Bernhard von Gudden and developed, as part of his doctoral thesis, a new staining technique which made it practical to count nerve cells. This was an important discovery, as it enabled scientists for the first time to establish whether any diseases led to the loss of nerve cells, as they could be now counted. Not only did his new staining technique (the Nissl stain) win Nissl a student prize for the best doctoral thesis but it had a far-reaching impact, as even today histologists and pathologists use 'Nissl staining' in the preparation of brain tissue for microscopy. Nissl arrived at the hospital shortly after Alzheimer, in 1889, and it seems they formed a friendship right from the off. Not only were they ideal work colleagues, one meticulous in patient note-taking (Alzheimer), the other meticulous in histology (Nissl), but they also complemented each other personally. Alzheimer was known as a bon vivant and a family man, whereas Nissl was a workaholic and eternal bachelor, but both developed a deep mutual friendship, which lasted until Alzheimer's death.

Alzheimer was keen to learn as much as possible from Nissl, and after Nissl's arrival, they improved and expanded the microscopy room on the first floor of the hospital. Since they were based at a psychiatric hospital, Alzheimer and Nissl were mainly interested in investigating the brain tissue of patients who had died in order to discover whether particular changes in their brain and nerve cells could explain their disease and symptoms. They used the latest staining techniques, including Nissl's own, on the brain tissue samples of the patients. Making nerve cells and their internal structures visible made it possible for the first time to observe any particular pathological changes in or around them, which could be related to the disease and symptoms that patients were having.

Over the next four years, Alzheimer and Nissl worked side-by-side and refined their clinical note-taking and microscopy techniques by learning from each other. In 1894, Nissl left Frankfurt for Heidelberg where he had accepted a position to work with Professor Emil Kraepelin, who was a colleague of his from his time in Munich. Nissl would go on to lead the histopathological institute at Heidelberg University until his death. Despite his move, Alzheimer and Nissl maintained a close correspondence by letter, exchanging patient and scientific information. Alzheimer had 'soaked up' as much knowledge as he could from the brilliance and expertise of Nissl, so that he was now himself an expert in microscopy and histopathology. The stage was therefore set for his ground-breaking encounter with Mrs Deter.

But before that disaster struck.

While 1901 may have been a professional high point for Alzheimer due to his fateful meeting with Mrs Deter, in personal terms it was to be a devastating time for the doctor. Earlier that year, Alzheimer's wife, Cecilie, died suddenly after a short illness, leaving the 37-year-old Alzheimer alone and with three small children (Gertrud, 6, Hans, 5, and Maria, 1) to care for. The love of his life was gone. Cecilie and Alzheimer had only married in 1895, but Alzheimer was besotted with her. (He never remarried and asked for his body to be buried in Celilie's grave in Frankfurt.) To help him with the children, Alzheimer's sister Elisabeth joined the household when Cecilie fell ill and stayed with the family until the children had grown up. The loss of Cecilie had a devastating effect on Alzheimer – not only had he lost his wife but he also missed his friend Nissl desperately. Nothing now held him in Frankfurt. To the dismay of Sioli, who was desperate to keep the trailblazing doctor, Alzheimer tended his resignation and followed Nissl to Heidelberg in 1902. Before he left, he went to see Mrs Deter to say a final goodbye and we can follow their, rather brief, conversation again, based on Alzheimer's notes:

Alzheimer: 'Good day, Mrs Deter.'
Deter: 'Ah well, just get away. I – cannot – speak.'

It is not perhaps the fondest farewell but we should realise that Mrs Deter's condition had by that point deteriorated significantly and she could recognise very few people. It is, therefore, likely that she did not recognise Alzheimer and simply wanted to be left in peace. Still, it's further evidence of Alzheimer's empathy towards his patients that he visited the wards for a final farewell before he left. Importantly, he also asked Sioli to keep him updated on Mrs Deter's symptoms and the progression of her disease.

4.
MUNICH VIA HEIDELBERG

Heidelberg has long been one of the most renowned medical schools in Germany. So it was a significant step up for Alzheimer's career to join this institution after coming from a 'backwater' such as Frankfurt, which had neither a medical school nor university at that time. Besides working with his friend and colleague Nissl, the other attraction for Alzheimer was to join Professor Emil Kraepelin's clinical and research group.

Kraepelin was at the time one of the leading figures in psychiatry. Not only did he write some of the first textbooks on psychiatric disease, but he was also a strong proponent of investigation into the biological causes of psychiatric diseases in the brain. It was the perfect clinical and scientific environment for Alzheimer, allowing him to continue his work with Nissl while at the same time learning from one of the leading figures in psychiatry. However, just when Alzheimer had settled in Heidelberg, Kraepelin accepted a position at Maximilian's University in Munich, where he moved his laboratory and clinic in 1903. Alzheimer decided to move again, following Kraepelin – but interestingly Nissl stayed in Heidelberg. This is curious since Alzheimer and Nissl were very good friends and one of the reasons why Alzheimer moved to Heidelberg was Nissl. It seems reasonable to speculate that Kraepelin must have made an enormous impression on Alzheimer, for him to follow Kraepelin to Munich.

Even more astonishing is the fact that there was no paid position for Alzheimer within Kraepelin's Munich clinic. Instead, Alzheimer worked – at least until 1906 – for free in Munich. He supported himself and his family, who moved with him, with the inheritance of his deceased wife. Cecilie was a widow of a Frankfurt diamond merchant when Alzheimer married her and so there was enough money to support them. The upside of not having a paid position was that Alzheimer could focus much more time on his beloved research and particularly the new microscopy suite he had installed in Munich. The other benefit was that he could spend more time with his children, to whom he was utterly devoted, according to the children's own accounts. Finally, the spare time also allowed him to maintain his correspondence with Sioli in Frankfurt and to receive updates on Mrs Deter. Sioli happily obliged and sent Alzheimer regular updates on her symptoms and well-being.

5.
1906

On 9 April 1906, Alzheimer received a phone call from Sioli. This was in itself a highly unusual event since the national telephone network in Germany was still in development and calling 'long-distance' from Frankfurt to Munich was not only difficult to arrange, but also exceedingly expensive. But Sioli had good reason to make this call as he told Alzheimer that Mrs Deter had died the day before. Alzheimer instantly asked if it would be possible to conduct a post-mortem on her and if Sioli could send him her brain for a histopathological investigation. He also asked for any clinical notes taken since his departure in 1902. Sioli arranged the postmortem on Mrs Deter and sent Alzheimer all the clinical notes from her case, along with her brain.

By modern standards, these actions were highly unethical. Neither Mrs Deter nor her family had consented to her autopsy or for her brain to be investigated further. We do not even know if they were informed of the procedures. But conducting such autopsies without consent was a common practice at the time. Only after World War 2, with the introduction of better and more stringent ethical guidelines for patient research, did asking for patients' consent become the norm.

Once Mrs Deter's brain and her clinical notes arrived in Munich, Alzheimer and two visiting doctors from Italy (Dr Perusini and Dr

Bonfiglio – whose role in detailing the first Alzheimer's disease case has been largely forgotten) began to prepare and investigate her brain. First, they carefully recorded its general appearance to the naked eye – the macroscopic investigation. This was followed by detailed histopathological investigations under the microscope, including various staining of the brain tissue. All three (Alzheimer, Perusini and Bonfiglio) independently agreed that the histopathological changes in Mrs Deter's brain were unique and had not been seen before. This caused some excitement and the three doctors continued their examination throughout the summer of 1906.

Because of these new findings, Alzheimer asked to speak at a regional conference for psychiatrists in Tübingen in Germany on 3 November 1906 – almost five years to the day after Alzheimer had met Mrs Deter for the first time. One can only imagine the doctor's great excitement as he presented her case and the histopathological findings to the audience, which consisted of some of the leading psychiatrists and neuroscientists of Germany. In typical Alzheimer fashion, he prepared his presentation for the conference meticulously and also produced photographic slides of the microscopic findings. It was truly innovative at the time to take photographs via the microscope and present them via a projector to an audience – a form of proto-PowerPoint. The amount of preparation was a further sign of how much importance Alzheimer placed on Mrs Deter's case. Finally, everything was set for his big day.

At Alzheimer's presentation in the afternoon, the chair, Dr Hoche, introduced him with the words: 'Next, Dr Alzheimer from Munich will report on: "A curious and serious disease of the brain cortex." Dr Alzheimer, the podium is yours.' At this, Alzheimer rose and started talking to his colleagues about the case of Mrs Deter. He first gave a detailed description of the disease history of Mr Deter. He reported that the initial problems that her husband reported were that Mrs Deter had started to forget information, misplacing kitchen items around the house and often becoming confused as to where she was and which day it was. Alzheimer then continued to report Mrs Deter's symptoms following her arrival at the clinic in Frankfurt and his interviews with her. He explained that when

1906

she arrived at the hospital her memory was already very poor and she had difficulties following conversations or writing sentences. He also emphasised that she was disorientated as to her exact whereabouts. Finally, Alzheimer reported that Mrs Deter died after four and a half years from an infection. This was probably caused by a decubitus ulcer (a pressure sore), a common infection for bedridden patients at that time.

The overview of her clinical presentation was then followed by the histopathological findings of her brain. Alzheimer had prepared four photographic slides showing the nerve cell changes in her brain, stressing that the nerve cells seemed to die due to an 'unknown metabolic substance'. On the first slide, he showed nerve cells which had 'disintegrated', leaving nerve cell fibres – or fibrils – behind, which he speculated must have been caused by the unknown substance. The second slide showed these fibrils in the different nerve cells of her brain, thereby demonstrating that many nerve cells were affected by the 'unknown metabolic substance'. He stated that 'between one quarter to three quarters of nerve cells' were affected by this new disease process. The third slide introduced a different change seen under the microscope of millet seed-sized spots in the upper layers of the brain (the brain cortex). Finally, on the last slide, he showed other different millet seed-sized spots and explained that they could be found outside of the disintegrated nerve cells and were filled with another unknown substance.

If these descriptions sound mysterious and puzzling to you, you are not alone. They are as puzzling to me and must have been to Alzheimer. However, such puzzlement is quite often how the process of scientific discovery works. A new scientific finding often starts with a simple description because there is no explanation – yet – for an observation.

What were these fibrils and 'millet seed'-sized spots?

No one knew at that time, including Alzheimer. All Alzheimer knew at this stage was that Mrs Deter, with the above-described symptoms and disease progression, had these changes in her brain.

Alzheimer concluded his presentation by saying that he found this to be a 'curious' disease process, of which there had been an increasing number of cases in the preceding few years. This statement again points to the fact that doctors must have seen many people similar to Mrs Deter prior to her case being detailed by Alzheimer.

If everyone already knew of such patients, what then was new here?

The key novelty lay in the microscopic/histological investigation of Mrs Deter's brain. This was truly ground-breaking at the time, since many doctors and scientists had presented the clinical presentation of psychiatric patients but very few had linked it to actual changes in the brains of those patients. It was because of these microscopic investigations that Mrs Deter was seen as the 'first' person to be diagnosed with Alzheimer's disease, even though many similar cases must have been seen or reported prior to this point by other clinicians.

Unsurprisingly, Alzheimer emphasised at the end of his presentation that more importance needed to be placed on such microscopic/histological changes in 'psychiatric diseases' to understand the disease better and help patients. He concluded that such a new approach would 'allow a more focused delineation of the disease' to improve the diagnosis and subsequent psychiatric treatment. Alzheimer fell silent at the lectern and eagerly awaited questions from the audience on his new and exciting findings.

But there were no questions.

He was met only by a wall of silence, while all eyes were on him. Finally, the chair of the session Dr Hoche said: 'Thank you, Dr Alzheimer, for your presentation. Apparently, there is no demand to discuss your findings.'

And with that Alzheimer was dismissed.

6.
THE TÜBINGEN AFTERMATH

As it so often happens, the true significance of a scientific discovery is not immediately apparent, even to experts.

Nevertheless, Alzheimer must have found the muted reception of his presentation utterly deflating. Let's remember that he had spent the previous five years following Mrs Deter's case while keeping meticulous notes and records. Furthermore, he had spent the last six months with his two Italian colleagues investigating her histopathological changes in detail, all apparently to no avail.

And to add insult to injury, the organising committee of the conference decided that based on the 'lack of questions and discussion', Alzheimer's case presentation would not be printed in an upcoming publication in a leading German psychiatric journal. In the days before the internet, the publication of selected findings of a scientific conference was a way to allow a wider audience access to ground-breaking research. Failing to be selected for a publication meant that Mrs Deter's case had not been deemed sufficiently important by the experts in the audience. It is also meant that the wider scientific and clinical community would not know about Alzheimer's new findings.

A devastating blow to Alzheimer.

Kraepelin, however, saved Mrs Deter's case from undeserved obscurity. He was insistent and convinced the committee that Mrs

Deter's case should be included in the publication of the conference findings. We can only speculate whether Kraepelin simply wanted to help his protegé or whether he saw the true significance of Mrs Deter's case. Regardless, his intervention would be pivotal in the 'discovery' of the disease, as without him the scientific article, consisting of Alzheimer's presentation at the Tübingen meeting, would have never been published in 1907. This very brief article – just shy of two pages long – is still regarded as the foundation stone for Alzheimer's disease.

Kraepelin's other critical intervention was to give the disease its name.

We already know that Kraepelin was a renowned psychiatrist at the time. Not only did he describe and classify some of the first psychiatric diseases, such as schizophrenia, but he had also written some of the first textbooks on psychiatry. For the first time, these textbooks were used to teach medical students and clinicians how to recognise and distinguish psychiatric diseases. However, the discoveries in psychiatry were moving so fast that Kraepelin realised that he needed to revise his original psychiatric textbook, published in 1904.

Throughout 1908, Kraepelin worked on this new textbook, the first volume of which was General Psychiatry, published in 1909, with the second volume, Clinical Psychiatry, following in 1910. Both volumes became instant classics for the clinical diagnosis and management of psychiatric diseases. Chapter VII of the second volume was dedicated to Senile and Presenile Insanity, which covered many psychiatric diseases in middle and older age. When writing Chapter VII, Kraepelin remembered Alzheimer's presentation in Tübingen in 1906. Kraepelin also remembered a patient, Johann Feigl, whom he saw in his clinic with Alzheimer in the spring of 1910 and who showed very similar symptoms to Mrs Deter's. Kraepelin described the case in great detail in his book. He also described the histological changes in the brain, which were analysed by Alzheimer, showing that Mr Feigl's brain had similar fibrils and plaques to those in Mrs Deter's brain. Kraepelin concluded the description of these cases in his book as follows: 'The clinical interpretation of this Alzheimer's disease is at the time [of writing] still unclear.'

This was the first time the name 'Alzheimer's disease' had been mentioned anywhere. Since Kraepelin's textbooks were widely distributed and read, it was the name which clinicians and researchers adopted to describe this new disease. It is, therefore, due to Kraepelin that we ascribe this type of dementia to Alois Alzheimer.

Ironically, Alzheimer himself was far more sceptical than Kraepelin as to whether those patients really had a different disease or simply were part of the – at the time – generally accepted 'dementia praecox'. But we should not forget that Kraepelin had a much broader overview of psychiatric diseases and he could clearly see that cases such as Mrs Deter's and Mr Feigl's were probably a new entity of disease. He therefore gave Alzheimer and his team more time and resources to find other such cases. This resulted in the publication of an article by Alzheimer and his colleague Perusini, which covered four cases of this new disease: Mrs Deter, Mr RM, Mrs BA and Mr SL in 1909. All of the cases showed very similar symptoms and all had the same histopathological changes in their brains. Along with Mr Feigl, these four people therefore provided the basis for future Alzheimer's disease diagnosis and research.

So, do we have only Alzheimer and Kraepelin to thank for the discovery of Alzheimer's disease? Not quite.

7.
FISCHER'S DISEASE

Have you ever heard of Fischer's disease?

No?

It's not really that surprising because Fischer's disease does not exist. However, Alzheimer's disease could have easily been called Fischer's disease or Alzheimer-Fischer disease because Dr Oskar Fischer published several cases of senile dementia at the same time in 1907, all of which featured the millet-sized plaques and fibrils in the brains that Alzheimer had observed. In fact, in 1910 Fischer published a scientific article running to more than 100 pages in which he detailed the histopathological changes in presenile and senile dementia. This far outweighed Alzheimer's observations and publications on the topic.

Given his substantial contribution, why have we never heard of Oskar Fischer before and who was he?

Oskar Fischer was born in 1876 in a small town northwest of Prague. He belonged to the German-speaking minority, living in what is now the Czech Republic. He studied medicine at the universities of Strasbourg, France (which taught in German at the time, as Strasbourg had been annexed by Germany from 1872 until 1918) and finally Prague. After obtaining his degree at the German University of Prague (Prague had two universities at the time – a German and a Czech one; the German university was incorporated into the Czech university in 1949), he worked at the Department of Pathological Anatomy and then the Department of Psychiatry until 1919. It was Fischer's time at the Department of Psychiatry that was to be pivotal to his career.

The Department of Psychiatry at the German university in Prague was at that time led by one of the leading behavioural neurologists – Professor Arnold Pick. Pick was a brilliant doctor and keen scientist. Pick is also a somewhat forgotten hero, as he discovered another kind of dementia – frontotemporal dementia. Pick was therefore keen for his department to become a leader in the study of presenile and senile dementia research and Fischer was part of this exciting institution.

Like Alzheimer, Fischer regarded microscopy and the new staining techniques as critical to making a breakthrough in research into presenile and senile psychiatric conditions. Before Alzheimer and Fischer, several people had already published observations of 'millet-sized' plaques in the brains of people with dementia. For example, in 1982 Dr Paul Blocq and Dr Georges Marinesco at the Salpetriere Hospital in Paris had already described these changes in the brains of people with senile dementia. Similarly, millet-sized plaques in the brains of people with senile dementia were described by both Dr Emil Redlich at the Second Psychiatric Clinic of the University of Vienna in 1898 and by Dr Koichi Miyake from the Psychiatric Clinic of Tokyo University in 1906.

What was new then in Alzheimer's and Fischer's findings?
The key to the definition of Alzheimer's disease was its observation of both plaques and fibrils in the brains of people with the disease. Only Alzheimer and Fischer described both these changes and should, rightly, be credited with the first description of the disease. In Part III of the book we will explore how both – plaques and fibrils – lead to Alzheimer's disease.

Instead of seeing Alzheimer and Fischer as lone geniuses, it's perhaps apposite here to quote Isaac Newton's observation that 'we can only see farther, because we are standing on the shoulders of giants'. In science the 'lone genius' is, in my opinion, a myth, since scientific 'discovery' is always based on the work of others. Both Alzheimer and Fischer were likely aware of the work of Blocq, Marinesco, Redlich and Miyake. However, it is both their observations of, and subsequent publications describing, both plaques and fibrils which represent their unique contribution to the definition of Alzheimer's disease.

But why does the disease not carry Fischer's name?

We can only speculate why this might be. One possibility is that Fischer thought that his plaque and fibril findings were specific only to one subtype of dementia – presbyophrenia. Presbyophrenia was a term used at that time to describe people with dementia who had an unusually elevated mood, were hyperactive and disoriented. Over time it became apparent that presbyophrenia might not actually be a different subtype of dementia and therefore the term was rarely used from the 1920s onwards. It's also possible scientists were unaware of Fischer's findings because his soon-to-be-outdated use of presbyophrenia as a term led scientists to dismiss his findings.

Another reason why we know the disease today as Alzheimer's disease is that Alois Alzheimer was a protegé of Kraepelin. This is not to diminish Alzheimer's contribution but the inclusion of Alzheimer's cases in Kraepelin's textbook made a considerable difference to the spread of the name, since Kraepelin's books were so widely distributed and translated. Still, I wonder whether Kraepelin was aware of Fischer's findings at the time of writing his revised textbook and, if so, why he decided not to refer to the new disease as Alzheimer-Fischer's disease?

The tragedy of Fischer's life lies not just in the fact that his scientific findings were overlooked, but that much of his working life proved to be a struggle. In 1919, he was denied tenure at the German university of Prague, which was highly surprising given that he had published significant scientific findings. He opened a private practice in Prague and continued to lecture as an 'adjunct professor'. Fischer's fate took an even darker turn in 1941 when he was arrested by the Gestapo and imprisoned at Theresienstadt (today Terezin, Czech Republic), which was both a ghetto and transit camp for Czech Jews being deported to concentration camps. Fischer was held in the 'small fortress' within the ghetto which operated separately and held mainly political prisoners and intellectuals. It is unclear why he was arrested in the first place. He had Jewish ancestry but was also highly political active in left-wing circles in the 1902s and 1930s in Prague. Fischer's brilliant life came to a brutal end when he died at Theresienstadt prison in 1942, aged 65. His seminal contributions to 'Alzheimer's disease' have been largely overlooked. It should carry his name as well.

8.
ALZHEIMER'S LEGACY

What about Alzheimer's legacy?

We now know that Fischer's legacy and his contribution to the discovery of Alzheimer's disease has been almost completely forgotten. Did Alzheimer at least achieve fame during his lifetime for the discovery of the new disease?

Far from it.

Despite his success in publishing his new cases and Kraepelin attributing his name to the new disease, Alzheimer was too busy to see the potential future impact of his discovery. The main reason for him being so busy was that in 1912 he had moved to Breslau (Wrocław in Poland today) to take up a professorship running his own clinic. Breslau was far away from Munich in Bavaria and hence Alzheimer was not as up-to-date on new scientific developments. Still, the position in Breslau was a dream come true for Alzheimer as he could design and run a department the way he finally wanted. His enthusiasm for the new position was evident when he arrived in Breslau and set up a new histological laboratory as well as a clinical service. He had finally achieved his dream of having his own clinic and laboratory, and more importantly, he could dedicate most of his time to his beloved microscopy. It should have been the beginning of a glorious time for Alzheimer, cementing his status as one of the leading psychiatrists in Germany at the time. But that glorious time was very short-lived.

Accounts from Kraepelin and other colleagues show that Alzheimer had already fallen ill during the move from Munich to Breslau and that after his arrival he spent some time in a sanatorium in Breslau to recover. Over the next three and a half years Alzheimer's health deteriorated further and he needed to take more frequent breaks. In particular, his heart seemed to have weakened, causing him shortness of breath. His kidneys seemed to have been in poor shape too. Despite his colleagues' best efforts, including those of Nissl and Kraepelin, who tried to support him remotely, Alzheimer died on 19 December 1915 at the age of 52.

As so often with people who die relatively young, his legacy was far from certain. While he always thought that Mrs Deter and the other cases were 'curious' and noteworthy, he did not pay any more attention to them than the other interesting cases he was meticulously following. Indeed, none of Alzheimer's obituaries at that time mention Alzheimer's disease or Mrs Deter. Instead, other psychiatric cases that Alzheimer reported were highlighted by his colleagues in the obituaries. One of life's ironies is that we do not know what we will be remembered for. Alzheimer himself probably would have been bewildered that his main legacy would arise from his fateful encounter with Mrs Deter.

Despite the occasional reports of cases, aside from Alzheimer's and Fischer's more seminal publications, little research into Alzheimer's disease would be published in the coming decades. It was only in the 1970s and 1980s that research into Alzheimer's disease was reinvigorated, probably for two reasons. The increasingly aged population in Europe and North America made the incidence of people with Alzheimer's disease in doctor's surgeries far more prevalent, increasing the need for better diagnosis, care and treatment. Secondly, science had moved on, improving our understanding of the typical symptoms of Alzheimer's disease and how it is caused in the brain. We are, therefore, now ready to explore the science behind Alzheimer's disease. First, let's have a quick recap.

PART 1.
SUMMARY

In this part of the book, we have covered the following aspects:

- The first person ever diagnosed with Alzheimer's disease was Mrs Auguste Deter.
- Mrs Deter was diagnosed by Dr Alois Alzheimer at a hospital in Frankfurt, Germany.
- It is very likely that many people before Mrs Deter had been seen and diagnosed by clinicians with similar symptoms.
- The critical difference in Mrs Deter's case was that Alois Alzheimer analysed her brain after her death and described specific changes, which are still considered the hallmarks of the disease today.
- It is the combination of Mrs Deter's symptoms and the changes in her brain which led to her being the first person to be diagnosed with Alzheimer's disease.
- Alzheimer presented the findings to colleagues at a scientific conference in Tübingen, Germany, but most of his contemporaries did not consider them noteworthy.

- The exception was Prof Emil Kraepelin, who became a champion of Alzheimer's findings and named the disease after him when writing a new textbook on old-age psychiatry.
- The forgotten person in the discovery of 'Alzheimer's disease' is Dr Oskar Fischer, who published similar findings around the same time as Alzheimer.
- However, Fischer faded into obscurity over the subsequent decades and therefore the disease carries Alzheimer's name and not Fischer's.
- Alzheimer himself did not live to see the legacy of his important findings, as he died soon afterwards and it took until the 1970s and 1980s for Alzheimer research to flourish.

PART 2.
MEMORY AND SPACE

Everybody knows what memory is.
But do we really?
Memorising a shopping list or the names of new acquaintances is a part of our daily life, but the true meaning of how integral memory is to one's self and personality only becomes obvious when we lose our memories, as happens with Alzheimer's disease. Memory is therefore not only important for our daily life but actually defines who we are.

The idea that memory is a central aspect of our personality had already been proposed in the 17th Century by the philosopher John Locke. He argued that our memory allows us to know that we are still the same person that we always have been. In essence, it means that we can remember being the same person that we have been for the previous decades, years, weeks, minutes or even seconds. Based on this, our life appears to us as one continuous entity and, more importantly, memory defines an integral part of 'our self'. We never even ask ourselves if we are still the same person that we always have been – we simply are. Of course, we have changed over our lifetime, by gaining more experience, and maybe a few kilos extra weight. However, we would never consider that the person who experienced events earlier in our life was a different person. It seems to be just a continuum of our life experience. But now, consider if we had lost, or were in the process of losing, our memories over the course of a lifetime. Other people might tell us about events which had happened but we would have no recollection of those events.

Does this mean we were present at those events or are they talking about a different person?

We would start to question ourselves and ultimately our identity.

Who am I really, if I can't remember who I was before?

This is not only a philosophical question but a very real scenario for people with Alzheimer's disease.

To gain an insight into the challenges that people with Alzheimer's disease face in dealing with memory loss, we need to first understand what memory is and how it works in the brain. Only then can we begin to understand the memory symptoms in Alzheimer's disease, and how best to support people experiencing those symptoms.

Let's start by exploring the different types of memory.

9.
'YOU ARE NOT MY TYPE'

It comes as a surprise to many that there are different types of memory, only some of which are affected by Alzheimer's disease. The most common type that people refer to when talking about memory is 'episodic memory'. Episodic memory, as the name suggests, is our memory of episodes or events in our life, such as remembering our wedding day or what we did last Sunday. In essence, episodic memory is our memory of personally experienced events. Episodic memory is the central memory function affected by Alzheimer's disease and therefore we will spend some time explaining how it works and how it is affected by the disease.

Other types of memory which are just as important as episodic memory but are less well known to most people. For example, 'semantic memory' is our knowledge of the facts pertaining to the world. In contrast to episodic memory, semantic memory is not bound to a specific event, but instead it is eventless information that we store in our brain. This sounds rather abstract but, simply put, semantic memory allows one to 'know' generic rules out there in the world that are not specific to an event. For example, we know that a cup is a cup, regardless of where we encounter it, its colour or its shape. Or we know that most birds have two legs, while most mammals have four legs, although, as with any rule, there are exceptions!

Semantic memory, then, is rule-based memory of the world and, not surprisingly, crucial to our everyday lives. However, we often take it as a given, since facts seem so simple to us. A cup is simply a cup. What is there to marvel at? But if we think about it, what defines a cup and how would you explain it to someone who had never seen or touched a cup? Is it simply 'a small bowl-shaped container for drinking from, typically having a handle', as the dictionaries tell us? How about non-bowl-shaped cups or those with no handles? We would still recognise it as a cup, even though it might not have all the features typically associated with a cup. But it is these features (bowl-shaped, handle, china) which we combine to learn the 'rules' of what is typical about objects or facts in our lives. Having seen many cups over the course of our lives, we simply 'know the rules' of a cup and we will not confuse it with something which has very similar features and uses – like a glass, for example. When people have problems with their semantic memory, they usually start to struggle when naming objects because they can't distinguish them from similar objects: is this really a cup or is it a glass? Fortunately, in the case of Alzheimer's disease, semantic memory remains relatively intact until the latter stages of the disease. Just remember when Alzheimer interviewed Mrs Deter and she struggled to remember any specific events of her life (episodic memory) but was able to tell him correctly the months of the year and the colours of objects (semantic memory). However, there is often confusion when a person reports to their doctor that they 'can't remember the words any more' because the missing memory could be episodic or semantic. In cases such as this, the doctor will then check whether the patient can remember the names of people they have encountered at recent events (episodic memory), for example; or whether they are unable to remember the names of everyday objects (semantic memory).

That we deal with both episodic and semantic memory in similar ways highlights how entwined our episodic and semantic memories are in our minds, even though they are quite different types of memories. We have explicit access to both semantic and episodic memories and they are therefore scientifically classified as 'explicit memory' types. 'Explicit' access means that we can recall the information of events or facts at any time, if we so wish.

Other memory functions which are not as easy to access deliberately are called 'implicit memories'. Implicit memories are largely used for any movements we carry out or skills we have learned. Imagine having to recall the movements for picking up a cup, for example. It would be very burdensome and slow us down. Instead, the brain bypasses our conscious mind and just 'does' the movement without us even thinking about it. Implicit memory is therefore used in many everyday skills. Even though we are very much consciously aware of how hard it is to learn a new skill, once we get the hang of it we do not think about it any more but just 'do it'. A classic example of such skill-based implicit memory is the ability to ride a bicycle. Learning to ride a bicycle is difficult, as it requires coordination of many body functions, including our muscles, our vision, our vestibular (balance) system and so on.

When we begin learning to ride a bike we have to focus on getting all these elements right. The first tries often lead to falls which provide us with important feedback on how to coordinate all the elements (movement, vision, balance, etc) correctly. It can take many hours to learn to cycle. However, once we have learned how to ride a bike, we do it without even thinking about what we need to do. Even more astonishingly, we will remember how to cycle even when we've taken a break from cycling for years or even decades. Once we have acquired them, implicit memories are stored in our brain for a long time and can be accessed over our lifetime. It is also very hard to explain to someone else the actions, since you do them instinctively. For example, it is hard to explain to anyone what you need to do to ride a bicycle. You just simply 'do it', so if you want to learn how to ride a bike, you just need to practise doing so. Other implicit memory functions are more related to conditioning, habit-building and priming, which are all connected to doing what you are doing automatically.

Implicit memories remain intact until the very late stages of Alzheimer's disease. Since those memory types are not as relevant to Alzheimer's disease, we won't discuss them any further – even though they are fascinating. Instead, we will focus on episodic memory, because it is a key memory type affected by Alzheimer's disease.

Next, we need to understand what factors influence our episodic memory and how we can have excellent episodic memory.

10.
THE ART OF MEMORY

In ancient times, excellent episodic memory was considered to be an 'art' and an important skill to possess. The main reason for this is that excellent episodic memory played a central role in everyday life and memorising 'by heart' was, for many people, the only way to remember information since paper was a scarce luxury. Episodic memory was used not only to remember things pertaining to everyday life, it was also used for storytelling and speeches.

Stories played an important role in entertaining and educating people and, before the written word, storytellers relied solely on their memory to recall them. To help them memorise these tales, ancient writers often used certain memory tricks. One 'trick' many epic, ancient stories – such as the Odyssey – employed was rhymes. Rhyming helps our brain remember information, as it is easy to 'fall into' the missing words if they rhyme with previous ones. The other 'trick' ancient stories employed was repetition. Repeating the same information over and over again allows us to consolidate (from Latin con = 'together' and solidare = 'firming up') the heard information.

I still remember, after reading Homer's Odyssey for the first time, thinking, 'I already know how that Odysseus is "swift-footed". Get on with the story!' But it was actually the repetition which helps me recall this phrase many decades later. Simply put, if we have heard a word or phrase many times we are more likely to remember it. Repetition is particularly useful when we can't write information down. Repeating information allows us to remember the information we hear and eventually to absorb it. It is not surprising that we can remember our favourite stories from childhood so well, since we wanted to hear them repeatedly. As a side-effect, our parents may remember them too, since they had to read them to us endless times. I can still recite by heart some stories I read to my children – Green Eggs And Ham, anyone?

A similar repetition effect can be found today in catchy songs. Catchy songs often repeat the same sentences or words over and over again, especially in their refrains. When you see these songs written down, it seems completely ridiculous how many times they repeat the same sentences or words, but it clearly works and our brains are highly receptive to this repetition. If you were given just the first part of a song or, even better, its refrain, you would easily remember, or even be able to sing the rest of the song. We now know scientifically that rhyming and repetition are key elements in our episodic memory, but we knew this already because a good part of our early education consists of repetition, with plenty of rhymes or songs to help us remember the alphabet, times tables or kings and queens.

However, if we take a step back, it is not actually the rhymes or repetitions themselves which help us remember information so well, but the fact that those rhymes or repetition bring order to the information we want to learn. By repeating the same information over and over, we therefore learn the order of that information and this order makes recalling this information more predictable. The order allows us to predict what we need to remember next. It is important to understand that ordering information is a key element for good memory and hence it has been used for centuries in our education. Ironically, however, the order of information is also the 'Achilles' heel' of our memory.

Anyone who has absorbed information by simply repeating it over and over again, so-called rote learning, knows that we can remember information quite well this way, as long as it is recalled in the same order. However, if the order is changed or interrupted, our recall can easily become confused. For example, try to recite a times table in reverse order or to recall the queens and kings of a royal lineage separately. It will be very difficult to do this using rote learning, as the order it imposes makes it much harder to recall the information differently. For truly excellent episodic memory we are missing two powerful key ingredients – objects and places.

To understand the importance of objects and places for excellent episodic memory we need to go back to the ancients again. They figured this out a long time ago. Today, only three ancient texts remain which outline the art of memory and give specific instructions on how to have the best memory: Cicero's De Oratoerm, Quintilian's Instituto Oratoria and an anonymous work called Ad C. Herennium Libri IV.

The first thing to notice is that these are not books about the art of memory but about the art of rhetoric. These days we think of rhetoric as the art of speaking but in ancient times it was thought of as the art of persuasion – consisting of five key aspects: invention, arrangement, style, delivery and memory. The persuasion of an audience could change people's minds and was therefore highly regarded in ancient times. It was the main method of convincing or persuading people of a cause, or simply educating or entertaining them – remember again the ancient storytellers and their tales. For ancient speakers, the only way to deliver an erudite and lengthy speech was to remember the whole speech or story by heart. This is fairly easy to do when a speech or story is less than five minutes long but now consider delivering a speech or story that runs to 45 minutes, one hour, two hours – or even longer, as the ancient speakers and storytellers often did. Professional storytellers could recite the whole Odyssey by heart.

There is no way of remembering such a story by rote learning. This was especially true of a speaker or storyteller in ancient times, who would have to deliver multiple speeches or stories on different occasions. This seems an impossible task unless the speaker has at least some written notes to help them through the speech or story, but the ancient speakers and storytellers managed just fine by using some 'secret' memory techniques.

11.
'SECRET' MEMORY TECHNIQUES

We do not know who invented these 'secret' memory techniques, but we do know that they existed a long time before Cicero and Quintilian wrote them down.

Even today you can find countless books selling these 'memory secrets', showing readers how they can improve their powers of recall. Titles such as Boost Your Memory or Limitless Memory might sound like they are based on modern science, but they largely rehash the ancient memory techniques mentioned by Cicero and Quintilian, who themselves quoted their predecessors. Nothing cutting-edge there then, but these are certainly well-proven methods. In essence, the key to extraordinary episodic memory is not only order (in the form of repetition or rhyming), but also the ability to combine it with visual information, such as objects and places.

We are highly visually-oriented beings and it's fair to say that the adage, 'A picture is worth a thousand words' applies very much to our mental processing. We can remember information much more easily when we have encountered it visually and our brain's visual system is large when compared to our other senses, such as hearing or touch. The same is true for most animals, which have excellent episodic memory, without having any written language. For example, animals recognise which food is safe to eat and where they can find reliable sources of water and sustenance. Evolutionarily speaking, it makes sense that our brain is wired in such a way that it is naturally drawn to objects and places, as they are vital for our survival and we can use this evolutionary predisposition to achieve excellent episodic memory.

The ancient authors knew that visual and spatial information is far more easily remembered than words or more abstract symbols. So the 'trick' the ancient orators used for remembering large amounts of information was to anchor our to-be-remembered information in real-world objects and space. This would make it far easier for us to remember it in the future. The final step in mastering episodic memory then is to combine our efficiency for remembering objects and places with the other critical memory ingredient which we've already encountered – order. Ordering the learned object and place information allows us to put the learned object and spatial information in a sequence so that we can remember it in a more structured way.

Now we have all the 'secret' ingredients, but how do we practically use these ingredients?

The key is to maximise our natural talent for remembering object and object information. Anyone can do this by simply placing the information we want to remember along a familiar route and use certain objects as markers for the things we want to remember.

Sounds a bit abstract, doesn't it?

Let's have a look at an example to understand how this method works.

Imagine we are starting a new job or joining a new club and want to remember the names of all the new people we encounter. There will be many people for us to meet and we will likely struggle to remember them all or even worse will get them confused. Believe me, I have been there; my wife still loves to tell friends that I introduced myself twice or even three times to the same person at the same event.

To learn all the names, we could simply print out a list of all the people's names and try to learn them. But this would be extremely time-consuming, especially if you need to remember many new people. We would probably also mix people up. 'Hang on, was Mr

Peters the small, thin man with the glasses or the thin man with the bald head?'

From this example, you can already see that we want to associate the name with a visual representation of that person. Any distinguishing features (glasses, bald head) are extremely useful for our memory to identify all these different people and their names.

Enter the ancient 'method of loci' (from Latin locus meaning place; literally meaning the 'method of places'), described by Cicero and Quintilian. For the method of loci to work we should distort or make people's features grotesque so that we can remember them and finally place those grotesque features along a route. The route can be a completely imaginary one or a real one. It could be, for example, rooms inside our house, a favourite walk of ours, our morning commute or any other route we know well. Once we have identified such a route we would populate it by 'placing' objects at different places along that route.

For example, our imaginary Mr Peters could be someone we want to remember on that route. But instead of placing Mr Peters as a name there, we want to place an object which reminds us of Mr Peters. The key is to choose a most absurd or grotesque object, as unusual or distinct objects are easier to remember. Let's say Mr Peters wears glasses and has a moustache. Along the route, we could place some giant glasses which we have to step past to continue on our way or, even better, we could have some thick 'Groucho Marx'-style glasses attached to a moustache. We will certainly remember Mr Peters after that.

However, we do not have to use the physical characteristics of people to place along the route, it could be any object which reminds us of that person. For example, another new colleague tells us that she loves playing golf. To remember her, we could place along the route a gate constructed out of some giant golf clubs or a miniature-golf course. With such information, we will not only remember her name but could also casually ask her the next time, 'So, what's your handicap?' She will likely be impressed that we remembered her hobby. Politicians and other people who have a lot of public interactions use such techniques to create some personal connection with the many people they meet. Even if we remember just one small detail of the person, it will likely make a big impression on them and put us in a good light. Maybe this all sounds a bit tedious, thinking about objects representing information you want to remember and finally placing them along routes. The good news is that our brain is automatically doing most of this for events in our life. Just think about it – an event like a birthday party usually takes place at a specific location with specific people or objects. Our brain, therefore, associates the people and objects (such as presents) with a particular place, say a restaurant where you celebrated a big birthday. If we recall this memory we can 'relive' the experience inside our head, seeing the place and everyone who was there in our inner eye.

The method of loci is therefore perfectly attuned to our brain's efficiency when learning and retrieving places and objects. But as we know from the above example it can get confusing when you have to remember a lot of very similar information, like the names of new people you have met. This is where the method of loci comes into its own. It is an extremely powerful way of remembering large sets of information with the least amount of effort.

To recall the information learned with loci, we just need to go along the route in our mind and the objects on the route will 'pop' into our head – hence its use by ancient speakers and storytellers. They simply used objects which reminded them of the next part of their speech or story along the routes. So a shipwreck would represent the next part of the story when Odysseus and his crewmates live on Calypso's island for seven years.

Clearly, the method of loci does not make sense when learning singular events (such as a wedding), as these are unique and our brain's normal memory capacity can easily cope with them. Instead, we use the method of loci for information which is very similar and hard to remember by using simple rote learning or our brain's natural ability. Just a word of caution: the method of loci is not perfect either and we need to avoid two mistakes: 1) we should not use the same route for different things we want to remember as they will interfere with each other. So keep the routes separate; 2) do not let the memory routes cross or overlap for some parts, as again this might interfere with our memory and we will get our different routes confused. Simply speaking, keep your memory routes separate.

I remember coming across this particular art of memorising when I was a student. I was reading the wonderful book by Frances Yates on the Italian philosopher Giordano Bruno and her own book on the art of memory. Having studied the brain, I was first struck by the fact that I really didn't know anything about the method of loci. No one had ever mentioned this technique in the lectures I attended and naturally I was sceptical that it was real, or perhaps I thought that it was just some mumbo-jumbo. But then my curiosity overcame my scepticism and I thought to myself, 'Why not give it a try?'

At the time, we were studying for a notoriously difficult exam on neuroanatomy which struck fear into students. The fear was justified as neuroanatomy is not only complex but requires remembering a large amount of information about different brain structures and nerves. Where did this nerve enter the skull and where did it cross over to the other side of the brain and which brain nucleus did it connect with? A nightmare to remember – but perfect for the method of loci.

So, I devised routes in my mind, along which I placed this information. One route, for example, was walking from the underground station at the university to my home. I simply placed the information I wanted to remember along the route. I found that particular street corners, street crossings and similar decision points that were perfect to place the information I wanted to remember. Each route was basically one nerve and simply following this route allowed me to remember all the information, as to where the nerve started, crossed, connected and so on.

Once I had learned the routes, I was startled by the efficiency of this method. It was very easy for me to remember the nerve information at any level and I could even go forward or backwards along the route to retrieve the information. My fellow students, meanwhile, struggled to learn all the neuroanatomy information by the usual rote learning method of repetition. I told quite a few of them about my 'new' method but to my knowledge no one used it, which I guess was natural since they did not want to abandon their tried and tested method of rote learning to remember this information. I still found the exam challenging but I could simply go along my route and remember the information the examiner asked. Over the years, I have used this method from time to time and it has not lost any of its power.

So, there we are, the 'secret ingredients' of excellent memory are order, repetition, objects and places. But there is one other element supporting episodic memory, which we need to understand in relation to the memory symptoms of Alzheimer's disease – attention.

12.
'ATTENTION!'

Despite my glorious neuroanatomical memory feats, my wife would tell you that my episodic memory is absolutely atrocious.

How is that possible?
Should I not have a perfect memory with my knowledge of the method of loci?

Unfortunately, she is absolutely right, I often do not remember things that I am meant to remember.

Does that mean that I have Alzheimer's disease or another neurological condition that affects my memory system in the brain?
The simple answer is, no.

The issue is more that I often do not pay attention to what my beloved wife says. This is not specific to her, as I have always lived in my head, or as one of my school teachers consolingly told my mother, 'He is just a bit of a dreamer.' To my wife's chagrin, that is certainly true and unsurprisingly it can drive people around you mad.

But this is nothing that I or we do consciously. Instead, our mind starts 'drifting away' and we begin to think about other things while we are busy with a separate activity or we have simply stopped listening to the person talking to us. After a while, we realise that we are no longer sure what we were meant to do or we have lost track of the conversation with the person sitting opposite us. This habit of drifting off or not paying attention can lead to some embarrassing moments but in my experience, it only really becomes an issue for older people and their fear of having Alzheimer's disease.

Why is that?
Over the years, I have heard so many people anxiously describing their memory problems like this: 'I go up the stairs and when I get to the top I can't remember why I went there' or 'I go to the other room and I can't remember what I wanted there.'

These so-called 'attentional memory problems' are generally regarded as being completely harmless and not a sign of Alzheimer's disease. The reason why such memory problems occur is that we become distracted or 'drift off' in our mind by thinking about something else. For example, while we are walking up the stairs to get something, we remember that we need to put the bins out later and 'wasn't that an interesting series that we watched last night? I certainly didn't see that ending coming.' Reaching the top step of the stairs, we wonder now, 'Why did I come here again?'

Another example of my 'memory problems' in everyday life is getting the children ready for the school run. There are always a million things to remember, such as lunch bags, sports bags, art homework and so on. Even worse, there is so little time to get this all ready, as we are running late and need to leave the house. So, it needs my complete focus. The downside is that because of my focus on the children, I get distracted from remembering the things I need for my own work, such as my swipe card or even my lunch sandwich. Again, my attention has been distracted by focusing on the children's things, while I forget my own day-to-day business. Admittedly, my long-suffering wife does much better in these kinds of circumstances and forgets less, whether that's regarding the children or herself. This highlights the fact that there are also individual differences and some people seem to be less susceptible to such attentional memory problems. Maybe it is just us dreamers?

Regardless, attentional memory problems are generally something we should not take too seriously, as we do with occasional 'memory slips'. Attentional memory problems are mostly nothing to worry about and are not a sign of Alzheimer's disease. We should therefore not worry when they occur. However, if our concerns persist, then it is always advisable to discuss it with our doctor, who can carry out more formal memory tests to determine whether we have the first symptoms of Alzheimer's disease.

We now know the key ingredients to good episodic memory. More importantly, we know that attentional memory problems should not worry us. Now let's explore where in the brain episodic memories are processed, as we will need this information to understand how these brain regions are affected by Alzheimer's disease and how that in turn causes the typical memory loss symptoms.

13.
'ENTER THE SEAHORSE'

The Papez circuit (named after Dr James Papez, a neuroanatomist, in the 1930s) is made up of different, inter-connected brain structures which contribute to our episodic memory formation, storage and retrieval. One of these brain structures is particularly important for episodic memory – the hippocampus.

What is the exact meaning of the word, hippocampus?
Hippocampus comes from Latin and means 'seahorse'. Why seahorse? The prosaic reason for this is that the 18th and 19th Century anatomists, who described most of the brain regions for the first time, named the brain structures according to how they looked. If you were to remove the hippocampus from the brain and examine it, you would be struck by how similar its shape is to that of a seahorse (as can be seen in the figure).

The hippocampus sits in a brain region called the temporal lobe, which is located just behind our temple, on each side of our head. This means that we actually have two hippocampi – one on the left side and one on the right side of the brain. Both sides have similar functions related to memory, but the left tends to remember verbal information while the right tends to remember visual information. Still, this is a subtle distinction and there is a lot of ongoing scientific debate as to how much the left and right hippocampus are specialised towards verbal and visual memory.

The hippocampus receives the visual information coming into our brain and is highly specialised in connecting object, place and time information. The hippocampus is, therefore, ideal for spatial navigation, as it allows us to connect the objects and places we need to find our way in the world and to know where we are at any point. Such object, place and time information is also critical for our episodic memory, since remembering events relies on knowing when, where and with whom an event took place. For example, 'Remember the day we went to see my parents and the car broke down on the way?' The hippocampus is therefore the perfect place for creating episodic memories as it combines object, place and time information, which as we know now from the method of loci are important factors for excellent episodic memory.

Based on this information, perhaps it should not come as such a surprise that the hippocampus is a key brain region affected by Alzheimer's disease. We will explore the exact mechanisms of what happens in the hippocampus and its neighbouring regions during the development of Alzheimer's disease in Part III of the book but for now, we simply need to accept that the hippocampus is important for Alzheimer's disease. Let's explore how Alzheimer's disease affects our episodic memory – and spatial navigation.

What we need to first understand is that the development of Alzheimer's disease in the hippocampus does not happen overnight but can take years, if not decades. The slow and gradual start of the disease means that the memory symptoms remain subtle at the beginning, as the majority of the nerve cells in the hippocampus continue to work normally. The episodic memory changes in Alzheimer's disease therefore occur gradually over time. In theory, the more Alzheimer's disease takes hold, the more nerve cells die in the hippocampus and in turn the worse our episodic memory should become. In reality, our hippocampus appears to be able to compensate for the loss of nerve cells for quite a long period, with our episodic memory remaining largely intact, but once the nerve cell loss is widespread there is often a significant loss in episodic memory. This occurs when Alzheimer's disease takes hold and the hippocampus becomes faulty, causing the typical memory symptoms found in Alzheimer's disease.

In theory, the slow, gradual start of the disease is a good thing as we can detect potential episodic memory changes early enough to treat them. However, in practice, the episodic memory changes can be so subtle they are hard to detect. Someone with early-stage Alzheimer's disease will tend to experience a gradual increase of occasional 'memory slips', such as failing to remember certain appointments, information or names. Clinicians often refer to this gradual development as being 'incipient', which literally means that the symptoms are 'creeping in'. This makes it not only more difficult to detect those episodic memory problems, but it's also harder to distinguish them from the sort of episodic memory problems which occur as part of our healthy ageing process.

It comes as a surprise to many people that our hippocampus changes with healthy ageing, without the presence of Alzheimer's disease. Indeed, our whole brain reaches its peak performance in our late teens and early 20s, as with our physical prowess. After reaching its peak performance in the 20s, our brain performance remains on a plateau before gently starting to shrink. This means that even in our late 30s and early 40s we see very subtle brain shrinkage, including to the hippocampus, which accelerates more when we are over 60 years of age.

The exact reason why our hippocampus and other brain regions shrink during ageing is not yet entirely clear. Some theories suggest that, as with other parts of our body, we are simply not as good at repairing or replacing cells in our body when we age, leading to a reduction in nerve cell populations and the shrinkage of various brain regions.

This age-related shrinkage of the hippocampus has no drastic impact on our episodic memory, but there are subtle changes. Most of us can attest that learning information becomes much harder, even when we are middle-aged, compared to when we were younger. I can certainly confirm this when I compare myself to my students who are able to 'soak up' information and retain it at an incredible speed, whereas I have to repeat the same information a few times. Retaining information is also harder and we are more susceptible to forgetting newly learned information. It is consequently very difficult to win against younger children in games like the aptly named Memory. Their level of recall for where the picture pairs in the memory game are hidden, for instance, is far more precise than ours. This lack of memory precision is exactly what is seen when the hippocampus ages.

Of course, the older we get the more we can compensate for this deficit with our larger overall knowledge of the world but learning new, detailed information becomes much harder. 'Teaching an old dog new tricks' is, therefore, not only a matter of attitude but also memory! The natural decline of episodic memory with ageing also has an important bearing on how we identify the earliest memory symptoms in Alzheimer's disease. Since episodic memory problems are considered a core symptom of Alzheimer's disease, anyone who thinks their memory has changed significantly and quickly should consider consulting their doctor. But because our brains naturally age, it can be harder to distinguish whether our memory problems are the first signs of Alzheimer's disease or are simply part of our healthy ageing. Are there any more specific episodic memory changes which can distinguish healthy ageing from Alzheimer's disease?

To answer that question, we need to understand a bit more about the neuroscientific basis of episodic memory and how it is affected by Alzheimer's disease.

14.
ENCODING, STORAGE, RETRIEVAL

Memory in the brain consists of three distinguishable but related processes: encoding, storage and retrieval. Encoding is simply the formation of a new memory. Many brain regions encode information but, as we've already seen, the hippocampus has a critical role in linking object, place and time information to create our episodic memories. Scientists are still exploring the exact mechanisms of how new episodic memories are formed in the hippocampus. However, we know that attention, repetition and the order of the information are important for the encoding of episodic memories. If those elements come together with information about object, place and time, we will encode a new episodic memory, which is then stored in our brain.

How our memories are actually stored is the least understood process in memory functioning. Various theories and models have been proposed over the years. We will only explore one dominant theory which has particular relevance for Alzheimer's disease – the Multiple Trace Theory. The Multiple Trace Theory states that over time the encoded information in the hippocampus is transferred to other brain regions for storage. This is important to understand, as it means that the hippocampus is not actually the place where our episodic memories are stored (something proposed by other theories). Instead, the Multiple Trace Theory argues that the hippocampus only stores multiple traces (or indexes) of our memories and not the actual episodic memory itself.

This sounds a bit abstract, so let's look at this theory a bit more.

According to the Multiple Trace Theory, we should think of the hippocampus as an indexing mechanism, which does not store the actual memory but rather the index of where the episodic memory can be found in the brain. As an analogy, let's liken the hippocampus to an internet search engine – the search engine does not store the actual website but we can find the website we are looking for by entering keywords. The keywords are similar to the indexes/traces the hippocampus has for each memory. If we enter the right keywords, we retrieve the right information. This is not an exact analogy and one key aspect missing is that the hippocampus might have links to multiple different details of the memory. Nonetheless, the more vividly we can remember an event, the higher the number of indexes or traces the brain will have for that memory. Not only can you remember where and when the event happened but also minute details such as colours, smells or other sensory experiences. The richest of episodic memories allow us to 'relive' an event, which we can play before our inner eye.

Finally, there is the retrieval process of accessing the stored information. Retrieval processes are highly reliant on so-called 'retrieval cues'. Simply put, the retrieval cue is the trigger that elicits the stored memory to be retrieved. Just imagine we want to remember some information. Most of the time, we would focus our attention on retrieving that information by simply asking ourselves – like, 'Was I meant to call John today?' The actual question we pose ourselves is therefore our retrieval cue. The question basically cues our hippocampus to retrieve the actual memory or to find the indexes for that memory and retrieve it. Such explicit retrieval, thinking about what you want to remember, is what commonly happens in everyday life. However, sometimes we simply can't remember the information, no matter how hard we try. Instead, we use other cues, such as our notes, diaries, or long-suffering partners to remind us what we were meant to remember.

On other occasions, we might not remember the required information and no one can help us, but then another time it seems to simply 'pop' into our mind while we are doing something completely different. In that case, our conscious mind has simply forgotten that we are looking for this information but our unconscious mind keeps looking for the information and finally delivers it. This can seem like magic because we are not consciously aware of the retrieval process at work.

This underlying process makes retrieval cues fascinating and we might have different cues for different memories. For example, smell can be a very strong retrieval cue for our episodic memories. We smell something and it transports us straight back to an event a long time ago or it simply evokes a good (or bad) feeling from the past. Music can be a great retrieval cue. Let's remember how catchy tunes can get stuck in our head and how we can remember the refrains of songs after years or decades. This also explains why people with dementia often can sing along to songs in the moderate or severe stages of their disease, even though they have lost access to most other memories.

'Access' is the keyword here. The fact that we can retrieve information even in advanced Alzheimer's disease highlights the fact that there is a big difference between losing an actual memory and 'only' losing access to a memory. This difference is key to understanding the memory problems in Alzheimer's disease.

15.
'IT'S ALL ABOUT ACCESS'

To reiterate the previous important point, episodic memory symptoms in Alzheimer's disease are mainly due to problems accessing memories and not their storage – at least in the early stages. To put it differently, people in the early stages of Alzheimer's disease still have their episodic memories stored in their brain but cannot access them any more.

Let's go back to our internet search engine analogy to explain this a bit further. We might be looking for a website but have forgotten its name and specific keywords to enter into the search engine or the search engine itself may be faulty. Under these circumstances, we are having difficulties accessing the website, even though it still exists somewhere. The information we enter in the search engine is therefore like a retrieval cue to find information in the brain. If we do not have the right retrieval cue or the memory index in the hippocampus itself is faulty, we will no longer have access to an episodic memory, even though it is still stored somewhere in the brain. These memory access problems explain the memory symptoms in the early stages of Alzheimer's disease.

Consequently, people with Alzheimer's disease can still have memories of events, such as what they did over the weekend but the faulty indexes will only allow diminished access to those memories. The people affected will merely retrieve the gist of those memories, such as 'we went to a restaurant', but they will have difficulty remembering details of that visit – what they had for lunch, for example. In this way, the faulty hippocampus indexes cause a reduction in details or vividness in the memories of people with early-stage Alzheimer's disease.

Does that mean that all memories remain intact in the brains of people with Alzheimer's disease and it is 'simply' a memory access problem?
For the early stages of the disease, this is largely true. However, in the latter stages of the disease, many other brain regions become affected by cell loss, which then causes the actual stored memories to be impaired and not only the indexes' access to them. Understanding this memory access problem explains most memory symptoms in the early stages of Alzheimer's disease. For example, the reason why people with Alzheimer's disease can remember certain things but not others, is due to the fact that the memory index remains intact for certain memories but not for others.

Another very common memory symptom which puzzles families is that people with Alzheimer's disease have very poor memory regarding recent events but a very good memory of events from long ago. The families often refer to this as 'living in the past', such as: 'His memory for what he did as a young man is perfectly fine but he can barely remember what he did last weekend.' Again, the key to understanding this 'living in the past' memory is the indexing mechanism of the hippocampus for accessing episodic memories.

One aspect of the indexing we have not mentioned yet is that the index of episodic memories gets stronger the more we remember a memory. Each time we retrieve a memory, the index/indexes for that memory are reactivated. This reactivation usually makes the memories stronger, although under certain circumstances they can also make them weaker. Since we have to retrieve a greater number of events from long ago on a greater number of occasions over the course of our lifetime, the indexes for those events are much stronger and more well-established than for more recent events. So we have much stronger indexes for older or more remote events, which we remember.

Once the hippocampus is impaired by Alzheimer's disease, it is, therefore the newer memories which are more vulnerable to the progress of the disease because they have been retrieved far fewer times, resulting in weaker indexing. On top of this, the indexes for recent events are either not arranged or are arranged in a faulty way since the hippocampus has already been damaged by Alzheimer's disease. These factors make it much more likely that people with Alzheimer's disease 'are living in the past', since they can remember those events much more clearly than recent events.

These same indexing problems for new events also cause people with Alzheimer's disease to ask the same questions over and over again, as they have literally 'no trace' of what they were told or experienced recently.

These problems indexing new memories can be extremely frustrating, of course, not only for the people with Alzheimer's disease but also for their families. Often the same questions are asked over and over again during the course of a day – occasionally interspersed with memories or stories from long ago, which have been told numerous times. It is important to remember that people with Alzheimer's disease do not do this on purpose. It is the faulty hippocampus which is to blame for them losing access to some, but not all, of their memories.

All of this should make it clear to carers and families that there is no point in telling a person with Alzheimer's disease that 'you have told them this already x times before'. They are not wilfully forgetting what you told them, instead their hippocampus has let them down and they can find no trace of what they have been told. Understanding this difference can lead to a significant change in how we deal with the memory problems that occur in Alzheimer's disease and improve the well-being of all concerned.

...

So far, we have mostly covered fairly abstract memory concepts in this part of the book. To put some 'life' into these concepts, let's now meet an imaginary person with Alzheimer's disease (Mrs

A) and her husband. I decided to use an imaginary person with Alzheimer's disease for two reasons:

1. To protect the identity of people I have seen with the disease over the years. Mrs A should be regarded as the archetype of a person with Alzheimer's disease we have seen as part of our research over the years.

2. To show all the typical symptoms of Alzheimer's disease. People present often with much more varied symptoms, so for explanatory purposes Mrs A is the 'textbook' person with Alzheimer's disease.

16.
THE IMAGINARY MRS A

Mrs A is a 72-year-old woman with a college degree and three grown-up children. Her husband, who accompanied her, greeted me with a very friendly: 'Good to see you again', which caused an almost imperceptible frown on Mrs A's face. Just missing a beat, she said: 'Indeed, nice to meet you again.' I asked them both to sit down and explained the procedure of the research appointment when I would ask them a number of questions.

When I had met them the year before I had asked them for a detailed history of how the disease started and what the current problems were. On the second time I saw them, I asked whether their symptoms had improved or worsened? Did they report the same problems as before or had other symptoms come to the fore?

Let's have a look at how Mrs A responded to my questions about her memory.

I turned towards Mrs A and asked: 'So, how has your year been?'
'Not bad, not bad.'
'Can you tell me anything, in particular, that happened, like holidays or family events?'
Turning towards her husband, she said: 'Um, we went on holiday, didn't we?'
That she needed to check already with her husband was a clear sign that her memory had worsened, as she could previously still

tell me the outline of recent events, such as holidays, without any help. Turning to an accompanying partner indicates that the person themselves cannot rely on their own memory but requires the help of someone else to remember what has been asked. Many clinicians call this 'the head turning sign' and it is usually the first indicator that something is awry with the memory of the person they are seeing.

Thankfully, Mr A remembered from last year to keep quiet as I wanted to find out how much Mrs A herself remembered, without his help. He simply shrugged at her.

Mrs A turned back to me and continued: 'Well, I think we went on holiday to France.'

'When was that then?' I asked.

'I am not sure, but it was hot, so it must have been in the summer.'

This is a good example of how people can often fill in the blanks in their memories without actually remembering the information. Since she remembered it was hot, she concluded it must have been in the summer, even though she did not remember the season or month.

'And did you visit any particular places?'

'Well, we went to the beach, to a place we have been before and just had a lovely time.'

'Anything else you remember from the holiday?'

'No, we just had a lovely time.'

Notice how I try to cue her into giving more details but I keep the cue generic ('Anything else you remember from the holiday?'), without giving her any information. I want to give her the chance to remember as many details as she can from the actual holiday.

So far we only know that Mr and Mrs A went on holiday in summer but we do not know when exactly, where exactly – France is a big country – for how long, if any other people joined them, where they visited or any highlights from the holiday. Any person with an intact memory would have given you a brief summary with at least some of this information. However, relying on only one memory is a bit tricky. Maybe Mrs A doesn't want to remember the holiday? I, therefore, changed tack to check for some other recent information.

'I also remember you have grandchildren. Did you see them a lot over the last year?' I asked.

'Oh yes, we saw them many times. It's so lovely to see them, even though I'm always exhausted when they leave. They're so full of energy,' she laughed.

'Can you remind me of their names and how old they are?'

'Well, there is Sophie who is 9, then there is George who is 6 and Harry is 4 – isn't he?' Mrs A turned again towards her husband.

This time he looked at her quizzically, raising an eyebrow, to which she quickly responded: 'Oh yes, I nearly forgot the newest member of our family. Jules was born two months ago.'

Asking for grandchildren's names and ages is another good way to get an insight into the state of recent memories. From her hesitant answer, we now know that she nearly forgot the most recently born grandchild, which can be explained by the fact that the memory index for this event is the most recent and therefore the weakest. Finally, I asked her to recall any public or news events that she remembered from the previous year. But again, Mrs A couldn't remember much information and if she could remember an event, she was only able to give me its outline, with few details.

Now I asked her husband to tell me about their holiday and their grandchildren. This allows us not only to corroborate the information but also provides a direct comparison between Mr and Mrs A's respective levels of information concerning those recent memories.

Turning to Mr A, I said: 'Could I ask you now to tell me what you did during your holiday and the names and ages of your grandchildren?'

Mr A then gave me a detailed account of their holiday in Italy (!), including the places and churches they visited – a lifelong passion of theirs. He also said their grandchildren were actually 11, 7, 5 and 2 years old, with the 'new-born' granddaughter being 18 months old.

It became clear that overall Mr and Mrs A were recalling the same recent events. However, the respective levels of details for those memories were quite different. Mrs A still remembered the events but seemed to get them confused, as she was not entirely sure which country they went to and when. Similarly, asking for the names and ages of the grandchildren elicited more generic information from Mrs A. She knew who they were and recognised them but the details, such as their ages, were much harder for her to recall. In contrast, Mr A was able to give detailed accounts of their travels and grandchildren, and if I had asked he could have recalled the whole itinerary of their holiday with detailed day-to-day description of things they did. As we already know, such missing details are mostly due to the faulty hippocampus in Alzheimer's disease. As the indexes of the memories are faulty, the memories are no longer arranged correctly or they are not retrieved in full detail, with only the gist of the memory remaining. So, instead of losing their memories completely, one finds that with Alzheimer's disease patients in the early disease stages, their memories become less detailed or 'sketchy' and sometimes jumbled up with similar events.

How about memories from long ago? Let's go back to Mrs A.

'Can you tell me a bit about your wedding day?' I asked.

'Oh gosh, this is so long ago but sure. We were married in a lovely little country church in Norfolk. It wasn't a fancy wedding like many people have these days, but we had a lovely time. I remember that my dress cost us a fortune and then I made it instantly dirty when stepping out of my dad's car at the church.' She laughed. 'My family was pretty much all there and quite large, while my husband's family was very small, so in the end, we mixed the sides in the church as it would have looked very odd to have a large crowd on the bride's side and very few on the groom's side. After the church, we had a reception in a nearby village hall. We had some lovely music being played by friends and most of the food was home-cooked. I can still see in my mind's eye, my mum in her kitchen cooking dishes until late in the night before the wedding day. It was simply a wonderful day!'

Notice the difference?

Mrs A has clear and vivid memories of her wedding day which happened many decades ago. We can imagine her stepping out of the car and how the church was arranged, just by what she is telling us. If prompted, I am sure she could have told me the food that was served and the music that was played at the wedding party – a detailed memory from many decades ago.

By contrast, she can't remember the details of last year's holiday or what she had done over the last two weeks. We already know that this is typical in Alzheimer's disease, with more remote memories being better preserved while recent memories are significantly less vivid. Also, notice the difference in her demeanour. She was quite hesitant about recent memories, checking or looking at her husband all the time, while she was completely relaxed talking about her wedding day, not needing his input at all.

This hesitancy in answering questions can often be another sign that the reported memory problems are not 'slips of memory', but a real change in the affected person's memory. Just imagine we had a slip of memory when talking to a friend. Nearly every person would laugh it off, thinking, 'How silly of me not to have remembered this.' Now imagine that we constantly had these types of memory slips. We would fail to remember different events or appointments, despite people telling us that we were at the event or that they had told us about the appointment. If this happens repeatedly, we can't simply laugh off the problem any more. Instead, we will become more careful or hesitant when interacting with people, as we can no longer be sure of what has happened.

It gives us a glimpse into how it must feel to have the faulty memory symptom of Alzheimer's disease. It is, therefore, no wonder that people experiencing such memory deficits become increasingly hesitant to say anything, as they can no longer be sure what has previously happened. Not only would this faulty memory be a problem for us but also severely disrupt our daily lives and the people around us. Forgetting recent details of a holiday or even the ages of our grandchildren might allow you to giggle and shake your head. But if people have problems remembering recent events, it becomes very hard to keep track of appointments, which bills to pay or even to switch the oven off after cooking. These changes to daily activities caused by faulty memory can often be the most distressing aspect of Alzheimer's disease as they affect everyday tasks that people have carried out for decades without any help.

Indeed, Mr A reported many instances where his wife had forgotten to switch off the oven or had put the dishes in the wrong cupboard. He also said that some days he felt like he was going crazy himself because she was asking the same questions over and over again – 'What are we doing today?', 'What are we having for lunch?' It is common to ask such questions maybe once or twice, but people with Alzheimer's disease can ask them numerous times each day, challenging the patience of the people around them. The best advice for people or families is to not argue back when a person with Alzheimer's disease repeatedly asks the same questions, as they simply cannot remember having asked that question due to their faulty hippocampus. Even if you explain to a person with Alzheimer's disease that you have told them five times already when lunch will be, they will have forgotten it in no time and will ask you again. Instead of confronting them, it is simply better to give the same answer again. Obviously, this is easier said than done, if you happen to live with a person with Alzheimer's disease who might ask the same questions 24/7.

Let's go back to Mrs A to explore the everyday changes and strategies one can use to deal with memory problems on a daily basis.

I asked: 'How are you with appointments or things you need to do during the day?'.

'I am still very good at it,' she replied.

'So, you don't have any problems with remembering appointments?' I asked her surprisedly.

'Of course I remember those, I might have dementia but I haven't lost my marbles – not yet at least,' she quipped back.

Producing a small diary from her handbag she added mischievously: 'Of course, this helps as well.'

As one can see with Mrs A, people in the early stages of Alzheimer's disease are quick at finding compensatory strategies, like her diary, to try to overcome their everyday memory difficulties. Having her diary and a calendar hanging prominently in the kitchen helped her to keep track of events and appointments, but without them, she would have difficulties remembering any scheduled events. Even having this diary and calendar would not overcome her difficulties in the long-term when she will likely forget to look into the notepad or calendar or even forget about them completely. Nevertheless, deciding together to use a diary and hang a large calendar in the kitchen allowed Mr and Mrs A to manage the memory symptoms in the short term. Instead of her having to repeatedly ask what's for lunch, they could write this information on the calendar and if she forgot, he could simply say, 'Why don't you look at the calendar?', instead of saying it again. Deciding on such compensatory strategies was a great way for the couple to deal with symptoms – instead of quarrelling over them – and to feel united in their fight against this disease. Even if such compensatory strategies will not help forever, they certainly have value.

Because of these 'compensatory' strategies, there does not appear to be much wrong with Mrs A, at least to the casual observer. She is simply a woman in her 70s who keeps a diary to keep track of her appointments. This impression is quite common in the early stages of Alzheimer's disease and clinicians often refer to this as 'appears normal to casual inspection'. It means, if one met Mrs A for the first time, one might not notice any problems at all with her. She would be simply a charming woman who had occasional memory lapses. However, people who know her well and have done for a long time would notice instantly that something was amiss, as her memory of recent events was somewhat sketchy and she would tell the same, old stories repeatedly. This would strike them as odd, as she would have had quite a good memory in general and would not have previously forgotten any significant events or that she told you the same story several times during the same visit.

'Oh, I nearly forgot to mention one thing', said Mr A. 'I have noticed that she is becoming increasingly disorientated and doesn't seem to be sure at times where she is. It is quite odd since we have lived in the same house for many years. But now she sometimes gets confused or even lost in our house and can't remember which door leads to where.'

'How about going somewhere outside, like the local shop or supermarket? Does she still do such brief shopping trips alone?' I asked.

'Alone? Definitely not. We had a real scare a while ago when she went on foot to the local shop just to buy some milk. But then she didn't return. I was getting really worried that something had happened to her and started calling people. In the end, we went searching for her with some neighbours and found her at the opposite end of the village. Since that time, I haven't let her out of my sight and we only leave the house together. It was simply too scary when she was suddenly gone. I've also bought a GPS tracker for her, which I put in her bag or pocket any time we leave the house, in case I can't find her.'

The last symptom mentioned by Mr A – spatial disorientation – is the other very common symptom in early Alzheimer's disease but is rarely reported, despite having potentially significant, if not fatal, consequences. Now let's explore this symptom in the following chapters.

17.
SPATIAL DISORIENTATION

Spatial disorientation is, along with episodic memory, a key symptom of Alzheimer's disease but while everyone seems to be worried about their memory when we get older, far fewer people have heard that disorientation can be also an early sign of Alzheimer's disease.

What is spatial disorientation?
Spatial disorientation refers to being unsure of where we are. We can all feel disoriented at times and unsure of our whereabouts. The perfect example is when we are travelling to a new place, such as when we go on a holiday. It is easy for us to 'lose our bearings' in a new town since we do not know its landmarks or layout. Only a map, navigation app or GPS device allows us to find our way around. It is completely normal to feel disoriented and lost in an unfamiliar environment. The difference in Alzheimer's disease is that people with the disease also get disoriented or lost in familiar environments, such as the town or neighbourhood where they have lived for many years. It can even occur in their own house, where they become lost and have to open all the doors to see which room is behind each door.

There is a perfect portrayal of such a scenario in the film 'Still Alice', about a woman with early onset dementia, played brilliantly by Julianne Moore. In one scene, the titular Alice needs to go to the toilet but no longer knows where the bathroom is, resulting in her having an incontinence accident. This is quite common and is very accurately depicted in the movie.

It's important to know that in Alzheimer's disease spatial disorientation can occur at any time, even in familiar locations, as people with Alzheimer's disease can lose their bearings and get disoriented. This disorientation is further exaggerated in unfamiliar places, where people with Alzheimer's disease are at a high risk of being disoriented or getting lost.

Despite spatial disorientation being a key feature of Alzheimer's disease, why is it so rarely mentioned?
There are several reasons why this might be the case. For example, the memory problems are simply more prominent and overshadow any other symptoms, such as spatial disorientation. In fact, we often see that people are more concerned about losing their memory than losing their way. Spatial disorientation is also a more 'private' symptom. Private in this sense means that the person with dementia might feel slightly disoriented at times but the people around the person might not notice the disorientation if it is not completely obvious. For example, the person with dementia hesitates just briefly on an outing but it is not clear whether they are unsure where they are or whether they have just thought of something. Finally, our brain can compensate for spatial disorientation symptoms for quite a long time. We can use other spatial orientation functions in the brain, which do not rely as much on the hippocampus to continue to find our way around, at least at the early stages of the disease.

Spatial disorientation tends to only become apparent to others when the disease has advanced further and we can no longer compensate for those deficits. In other words, the spatial disorientation is always there but our brain can compensate for this loss for a while. Scientific evidence has shown that 70% of people with dementia have had at least one episode that involved getting lost, revealing again how spatial disorientation is an integral part of the condition.

Spatial disorientation is often considered to be a 'knock-on' effect of the pervasive memory problems in Alzheimer's disease – that people with the disease can't remember the locations or how to find their way from one place to another. While it is true that memory contributes to spatial disorientation since both rely heavily on the hippocampus, the more significant contribution comes from changes to the actual spatial navigation system in the brain caused by Alzheimer's disease.

To understand how the spatial navigation system in the brain is affected by Alzheimer's disease we need to first understand how the spatial navigation system works in healthy brains. We already know that the hippocampus is important for episodic memory but it is also a key region for our spatial navigation functions. The central importance of the hippocampus for navigation only became apparent in the 1970s when research studies found that the different nerve cells in the hippocampus were only active in certain locations. Hence we call these nerve cells 'place cells', because they seem to record the place where we are. In the 2000s another type of navigation-specific nerve cells called grid cells were discovered and these have become critical to our understanding of Alzheimer's disease. Grid cells are located in the entorhinal cortex, next to the hippocampus. Grid cells do not record our location, like place cells, instead they record and measure our movement through space.

The discovery of place and grid cells meant that, for the first time, scientists could establish how the brain keeps track of where we are and move, in a similar way to how the GPS in our car works. But spatial navigation functions in the brain are not only confined to the hippocampus and entorhinal cortex. There is a whole network of brain regions involved in our spatial navigation processes. However, the reason I am highlighting the hippocampus and entorhinal cortex is because both are the earliest regions affected by the proteins in Alzheimer's disease. Since those regions are critical for our spatial navigation, it also explains why people with Alzheimer's disease become disorientated or even lost.

We already know that the hippocampus and its neighbouring regions are affected by Alzheimer's disease and that this can cause significant memory problems. But since we also know that these regions are important for our spatial navigation and orientation, it makes sense that people with Alzheimer's disease should show problems with navigation and orientation. A more surprising fact is that such spatial navigation/orientation symptoms might actually appear before people with Alzheimer's disease develop memory problems. The reason for this is that the entorhinal cortex, which is responsible for the grid cell functioning in navigation, is one of the first regions affected by Alzheimer's disease, even before the hippocampus is affected. In Part III of the book, we will explore in more detail how these regions are affected by the disease but we know that the proteins responsible for Alzheimer's disease (amyloid and tau) accumulate in the entorhinal cortex before the accumulation spreads to the adjacent hippocampus. This means that the nerve cells in the entorhinal cortex and particularly the grid cells will likely be affected before the cells responsible for the indexes of our memories in the hippocampus.

Does this mean we should pay more attention to the entorhinal cortex and spatial navigation functioning in the early stages of Alzheimer's disease, instead of focusing on the hippocampus and episodic memory?
It's a good question, and one for which we currently have no answer. One reason for the focus on the hippocampus and episodic memory is that these changes have been known and recorded since Alois Alzheimer's and Oskar Fischer's publications in 1907. Contrast this with the entorhinal cortex findings, which only emerged in the 2000s. This might explain why we still require a shift from episodic memory to spatial navigation. In fact, very few cognitive diagnostic assessments for Alzheimer's disease so far have tested spatial navigation and orientation. Time will tell whether spatial disorientation becomes a future key diagnostic criterion for Alzheimer's disease and one which might facilitate the identification of the condition even before people develop memory symptoms.

Regardless of this, it is clear that spatial disorientation causes significant everyday problems for people with Alzheimer's disease and their families. In particular, getting lost is an overlooked but potentially dangerous symptom.

18.
'LOST IN SPACE'

Getting lost or going missing is a common problem in people with Alzheimer's disease. In the UK, around 40,000 people with Alzheimer's disease get lost every year. The vast majority of people are safely found shortly after getting lost. However, every year there are a number of tragic cases when someone with Alzheimer's disease gets lost so significantly that they either come to harm or even die from exposure.

One such person was George Herbert, who went missing in the English county of Norfolk and sadly died. But his death was not in vain, as the subsequent inquiry recommended the establishment of a protocol to help police and search and rescue groups to find people with dementia faster or prevent them from getting lost in the first place. The Herbert Protocol is now used by police throughout the UK. The person with dementia and their family complete a form which includes basic information about them, as well as routes they would have commonly taken and places they often went to, such as former places of work or residency. To safeguard against someone with dementia getting lost, the family gives the Herbert Protocol form to the attending police or search and rescue services, and this will help to locate the person with dementia.

What else can people do to prepare for such episodes?
There are a few key aspects which are important to understand regarding when and how people with Alzheimer's become lost, and this will help them and their carers and families to be better prepared.

The first key thing to understand is that the first time this happens is often completely unexpected. Since the spatial disorientation in the person with Alzheimer's disease is often not known to the carer or family, they assume that the person is still OK to go out by themselves. We already know that the spatial disorientation is present right from the beginning of the disease and people with Alzheimer's can compensate for this for quite a while. However, once the compensation no longer works, they can get lost even on routine outings, such as taking the dog for a walk or going to the local shop. As these sorts of outing make up the bulk of our day-to-day lives, they are also the most common time for people with Alzheimer's disease to get lost.

The other main occurrence is referred to as 'waiting in place'. This occurrence is, in fact, a combination of issues arising from the memory and spatial orientation deficits found in people with Alzheimer's disease. Waiting in place situations are often those during which the person with Alzheimer's disease and their carer are out together and the carer asks the person to wait somewhere while they do a quick errand, such as waiting in front of a shop while the carer quickly goes into the shop to buy a few items. In this situation, the person with Alzheimer's disease can actually forget what they have been told to do. Instead, they wander off from the place where they were meant to wait. This would not necessarily be a problem, were it not for the fact that the person's spatial disorientation leads them to become lost. At the same time, the carer comes out of the shop and their loved one has vanished and cannot be found nearby.

Getting lost on routine outings and waiting in place account for the vast majority of missing cases in people with Alzheimer's disease and it is, therefore, well worth being aware of them. There are two other instances where a person with Alzheimer's disease might get lost, even though they are much rarer. The first instance is prompted by agitation, such as being upset or angry with the carer or themselves. This agitation leads the person in question wanting to leave the house. The other rarer occurrence is when the carer is asleep or not in close proximity to the person with Alzheimer's disease and they walk out of the house because they feel they have to visit a place. For example, people with Alzheimer's disease often get confused in the later stages that they want to go 'home', meaning their home as a child, or that they need to go to work, even though they have long been retired. The key difference between the more common and rarer occurrences of getting lost is that the common occurrences happen when people are already outdoors, whereas the rarer occurrences all happen when people leave their 'home.'

What should we do when an event such as this has occurred?
First of all, the key thing is to keep calm. The vast majority of people with dementia are found within a 3km radius of their residence and within one hour. The police and search and rescue services call this the 'golden hour', as people found within the first hour of going missing are less likely to come to harm. So if someone is missing we need to act fast and should always contact the police, even if it's quite possible that the person will be found within five minutes – for example in a neighbour's garden.

In the UK, completing a Herbert Protocol should be a priority for anyone with a diagnosis of Alzheimer's disease and their family before such events can potentially occur. Simply fill in the form and keep it in a safe but accessible place in case a person with Alzheimer's disease gets lost. Police in other countries also have various systems and mechanisms in place to deal with missing people with dementia. In the US, for example, the Silver Alert public notification system broadcasts information about such missing persons in order to help in locating them.

Is there a way to prevent such events occurring?
There are a number, but none entirely eliminate the risk. Various technological measures such as door alarms or sensors can let the carer know that the person in question is leaving the house. They are a good complementary way of 'keeping an eye' on the person with Alzheimer's disease and giving the carer more peace of mind.

Door alarms are standard in nearly all dementia hospital wards or care homes, and unsurprisingly the rates of getting lost are diminishingly small for hospitals and care homes. It is well worth considering installing such alarms/sensors at home.

The best option for preventing getting-lost events during an outing is a GPS device. Dementia-specific GPS devices can be either installed as apps on a smartphone or in the form of little boxes which can be put in bags or pockets. GPS devices allow the carer or family to locate the position of the person with Alzheimer's disease. For some people, this might be too much of an intrusion into their privacy, but it provides peace of mind to carers or family members as we always know where the person with Alzheimer's disease is, providing they wear the tracker.

This is key. The person with Alzheimer's needs to carry the GPS device in order for it to be effective. If they leave their mobile phone or tracker at home, it will not work. Similarly, if they remove it, drop or throw away the GPS device while outside, they cannot be located. Despite these shortcomings, GPS devices can be extremely helpful because they not only help to locate the person with Alzheimer's but also allow you to create so-called geofences. A geofence is an area we define as somewhere the person carrying a GPS device can roam freely. However, once they leave this area and breach this virtual geofence, an alarm or text message is sent to the carer, alerting them that they have left their defined area. This more sophisticated approach gives the person with Alzheimer's disease more freedom and autonomy while also allowing their carers to continue safeguarding them.

Getting-lost episodes not only have an impact on the person with dementia, but also on their carers and families. The scientific evidence shows that carers of people with dementia who go missing repeatedly have much higher stress and care burden levels compared to other carers. Repeated getting lost episodes also lead to a sevenfold increase in care home placement of the person with dementia, further reducing their autonomy and independence. It is therefore important to strike a balance between the safety and the autonomy of the person with dementia when considering how to deal with this type of situation.

As we can see then, spatial disorientation in Alzheimer's disease is not only an important factor when it comes to detecting significant cognitive changes as early as possible, but it also affects safeguarding for people with the disease, their carers, and their families.

PART 2.
SUMMARY

In this part of the book, we have covered the following aspects:

- There are different types of memory of which episodic memory, the memory for personally experienced events, is the most deeply affected in the early stages of Alzheimer's disease.
- Episodic memory relies on combining object, place and time (what, where, when) information to create vivid memories.
- The hippocampus is a key brain region involved in our episodic memory. The hippocampus does not store the actual memories but creates indexes for all the details of the memories.
- The hippocampus shrinks with healthy ageing and hence our memory, in general, worsens as we get older.
- In Alzheimer's disease, the hippocampus is directly affected by the disease which causes nerve cells to die and therefore affects the memory indexes.
- Faulty memory indexes cause poorer memory – that is, a memory with fewer event details. Often only the gist of the memory is retained while many details cannot be retrieved.

- In the early stages of Alzheimer's, we have more of a memory access problem, that is we can't access the memories in full detail any more due to the faulty indexes in the hippocampus.
- The indexes for older memories are much stronger than for newer memories. Therefore, people can often retrieve memories from long ago in much greater detail than recent events, which is why people with Alzheimer's disease often seem to be 'living in the past.'
- Spatial disorientation is a common but under-recognised symptom in early Alzheimer's disease. Spatial disorientation is caused by proteins accumulating in the brain spatial navigation network, which partly overlaps with the memory network.
- Some parts of the spatial orientation network appear to be the first affected by Alzheimer's disease, even before the memory network.
- Spatial disorientation can therefore be potentially the earliest symptom of Alzheimer's disease, even before memory problems.
- Spatial disorientation also has significant everyday implications with people with Alzheimer's disease as they can potentially get lost and come to harm. This has implications for balancing autonomy and safeguarding people with Alzheimer's disease.

PART 3.
AMYLOID AND TAU

Let's do a quick recap on what we know so far about Alzheimer's disease before we go deeper into how Alzheimer's disease develops within our brain cells.

Part I of the book introduced some key players in the history of Alzheimer's disease, particularly Dr Alois Alzheimer and Mrs Auguste Deter. We now know that cases of Alzheimer's disease had been described before Alzheimer's time. The key advance he made, however, was to identify not only the observable changes in Mrs Deter's memory but also the specific physical changes in her brain that were associated with her disease. Microscopy was key to spotting two 'unknown substances' in Mrs Deter's brain, which Alzheimer concluded must be related to her symptoms and disease. It was these protein accumulations which Alzheimer reported for the first time at the Tübingen conference, to an underwhelming response.

In Part II of the book, we discovered that protein build-up causes the typical memory problems we see in Alzheimer's disease. Proteins accumulating in the hippocampus cause its indexing mechanism to become faulty, which leads to problems accessing memories. Recent memories are particularly vulnerable to this protein accumulation and hence people with Alzheimer's disease have particular difficulties accessing those recent memories, while their remote memories stay largely intact. It seems as if they are 'living more in the past'.

What then are these protein build-ups or 'unknown substances', as Alzheimer (and Fisher) described them? And how are they related to the memory problems inherent in Alzheimer's disease?

Let's start off with what proteins actually are and why they are important for our bodies.

19.
GLORIOUS PROTEINS

Proteins are large, complex molecules which consist of amino acids. Amino acids are often called the 'building blocks of life.' The sequence and number of amino acids in each protein determines its structure. This structure is important because it determines which role proteins carry out in our body. One can compare protein structures to keys. The key needs to fit perfectly in the lock to open the door. Similarly, the structure of proteins define how they dock onto or are embedded into different cell structures to enable the cell's function. Within the cells of our body, proteins play several different roles in the structure, function and regulation of our tissues and organs.

It should come as no surprise, then, that when proteins 'go wrong' it can cause havoc in our bodies. The same applies to our brain, where changes to proteins can cause significant problems and are the cause of most neurodegenerative diseases, such as Alzheimer's disease, Parkinson's disease, Huntington's disease and Motor Neurone Disease. Despite their various symptoms being very different from each other, all these diseases have one thing in common: they are so-called 'proteinopathies'. Proteinopathy is a disease that is linked to changes to proteins.

In some proteinopathies the proteins are not correctly built. In others the proteins are not correctly dismantled.

Other proteinopathies cause proteins to be 'overexpressed' or 'underexpressed', meaning that either too much or too little of the protein is produced, causing significant changes in the brain. It is important to understand that there is not only one process affecting the malfunctioning of proteins. Instead, a variety of malfunctions, such as being incorrectly built, dismantled, or over/underexpressed results in different proteinopathies.

There is one more point to understand regarding proteinopathies – different proteins cause different dementias. For example, Alzheimer's disease is caused by a different proteinopathy from Frontotemporal Dementia or Dementia with Lewy Bodies. And to make things more complicated, some forms of dementia can have very different symptoms that are caused by the same proteinopathy. Is all this complexity making your head spin?

Do not worry, even experts in the field struggle to distinguish between the different types of dementia and their proteinopathies. For now, it is enough to understand that the underlying malfunctioning of proteins may differ but it can also overlap among different types of dementia. This is not only theoretically interesting, it has considerable implications for developing treatments for such proteinopathies, which we are going to explore later in this book.

So, which proteinopathies cause Alzheimer's disease?

In Alzheimer's disease, the key proteinopathies are caused by two proteins – amyloid and tau.

20.
AMYLOID

Amyloid is a key protein involved in Alzheimer's disease. Let's cast our minds back to 1906 when Alois Alzheimer reported seeing bundles or 'plaques' of proteins outside of nerve cells. He called them millet seed-sized spots filled 'with an unknown substance'. Only in the 1960s, with the advent of electron microscopy, did we understand that these spots were plaques filled with filaments of protein. However, what the exact protein was and how it came to aggregate in these plaques remained unclear at that time. It took another 20 years until several research groups identified the protein as amyloid. More specifically, beta-amyloid, which is one particular type of amyloid. One way to imagine these amyloid plaques is to compare them to the balls of dust and hair that we find often under our sofas. Similarly, amyloid plaques are 'balls' of randomly accumulated beta-amyloid.

Why would proteins accumulate in balls/plaques outside of nerve cells? What is their function? Well, these plaques are actually waste products from the nerve cells. Just as balls of dust gather behind our sofas, so the amyloid has accumulated outside of nerve cells.

But why would this waste accumulate outside of nerve cells? Is it not bad for the body to let such waste accumulate?
It is indeed not good for the body to let such protein waste accumulate and our brain has several mechanisms to get rid of excess beta-amyloid. However, once beta-amyloid forms plaques it

is an arduous process for the brain to break them down and remove them.

Why do these amyloid plaques occur more often as we age, potentially going on to cause Alzheimer's disease?
This is a great question. Theoretically, amyloid plaques can form at any age if there is too much beta-amyloid in the brain. However, when we are younger, our brain is far more efficient in removing the amyloid. This changes as we age, with our brain starting to slow down the disposal of amyloid plaques. The exact reasons why this clearing slows down with ageing are still being investigated. Regardless, it means that as we age we have an increased risk of developing amyloid plaques in the brain.

Should it not then follow that everyone gets Alzheimer's disease at some stage in their later years, since amyloid plaques keep accumulating over time?
To answer this question, we need to take a step back and explore how beta-amyloid and amyloid plaques are formed in the first place. This will help us to understand why some people can have amyloid plaques in the brain without actually developing Alzheimer's disease.

21.
BETA-AMYLOID FORMATION

To understand how beta-amyloid is formed we need to look at the genes which influence amyloid's formation, activity and disassembly. One is particularly important: Amyloid Precursor Protein (APP). As the name makes clear, this particular gene is the blueprint for the Amyloid Precursor Protein, which is a large protein sitting in the membranes of cells where it has multiple functions, including the growth and repair of nerve cells. The Amyloid Precursor Protein leads something of a Dr Jekyll and Mr Hyde existence. As Dr Jekyll, it is vital for growing and repairing healthy nerve cells. However, it also has its darker Mr Hyde side. When it gets broken up, one part of the Amyloid Precursor Protein can create beta-amyloid.

How does this happen?
All proteins have a lifespan and once this time is up, they get broken down by another class of proteins called the secretases. Imagine secretases as scissors snipping other proteins into pieces to disassemble them. This 'snipping' is not random, but highly specific. Each secretase cuts each protein at highly specific places. The resulting smaller, residual parts of the protein can then be either used for other purposes in the cell or be processed as waste material.

Three secretases (alpha-secretase, beta-secretase and gamma-secretase) exist to disassemble Amyloid Precursor Protein. Of most importance in Alzheimer's disease is the beta-secretase which snips the Amyloid Precursor Protein into two components: an Amyloid Precursor Protein Beta part and beta stub. The important beta stub is further cut by the gamma-secretase, creating beta-amyloid.

After the secretases have done their snipping work, the beta-amyloid initially remains in the cell membrane, but since it now has no function it is soon ejected out of the nerve cell as waste material. Outside the nerve cell, beta-amyloid awaits its fate as waste material to be transported or flushed out of the brain. As when we put out the rubbish for collection, the beta-amyloid waits to be taken away via the corticospinal fluid, or the blood, and to be further broken down and recycled by the body. The figure shows how the processes differ between the normal and abnormal cutting of the amyloid precursor protein.

We know already that this flushing out of the beta-amyloid can slow down during ageing. This means that over time beta-amyloid accumulates outside of nerve cells. When this happens, the beta-amyloid molecules can start to glue or clump together, forming sheets of beta-amyloid, a bit like glueing lasagne sheets together lengthwise. These sheets look like strings under the microscope. The scientific term for such strings is fibrils. Although the sheets are not yet amyloid plaques, they are referred to as amyloid fibrils or oligomers.

The two most common types of beta-amyloid fibrils are beta-amyloid 40 and beta-amyloid 42. The 40 and 42 refer to the number of proteins in each beta-amyloid fibril. The important difference between beta-amyloid 40 and 42, is simply that beta-amyloid 42 increases our risk of developing Alzheimer's disease. (The exact reason why is still being investigated by scientists.) This is because as the number of beta-amyloid 42 fibrils between the nerve cells increases, they start to form amyloid plaques which we can see under the microscope. When we try to remove or recycle a big sticky mess it is much harder and takes far longer than when we deal with each component separately. It is the same for the brain and so we find that amyloid plaques can stay there for a long time, until they are broken down. Some plaques might never be broken down and will stay with us throughout our lives. This explains why we have an increasing amount of amyloid plaques in the brain the older we get, as there is simply more time for them to accumulate and they are harder to get rid of.

It should become clear now that the accumulation of amyloid plaques is a slow process, as it takes a long time for beta-amyloid proteins to accumulate and clump together into amyloid fibrils and, subsequently, for the fibrils to form amyloid plaques. It also explains why the development of Alzheimer's disease can take such a long time – years or even decades – as the accumulation of amyloid is a slow process. It further explains why Alzheimer's disease is seen as 'incipient', as the amyloid accumulation creeps into our brain over time and we only notice that we have too much amyloid in our brain when we start developing symptoms.

Unfortunately, by the time we develop symptoms our brain is often already quite full of amyloid fibrils and plaques. But this is not all bad news. From a prevention or treatment perspective, to have such a slow accumulation means that there is quite a large 'treatment window'. This large treatment window means that we have potentially years, if not decades, to reduce or stop the accumulation of beta-amyloid in our brain. All we must do then is to remove or reduce the beta-amyloid from the brain and we can treat or even cure Alzheimer's disease.

This was the idea behind the Amyloid Cascade Hypothesis from 1990, which stated – quite reasonably and brilliantly at the time – that the more beta-amyloid we had in our brain, the more likely we were to develop Alzheimer's disease. Quite simply, the theory stated that beta-amyloid in the brain triggers a cascade which eventually leads to Alzheimer's disease.

This hypothesis became the key to Alzheimer's disease research and drug trials for many decades. However, it turned out that this hypothesis was too simplistic as there was another mysterious substance at work.

22.
'LOCATION, LOCATION, LOCATION... AND TIMING'

In the 2000s, a puzzling finding emerged showing that the hippocampus – the region we most associate with Alzheimer's disease and its memory symptoms – seems to be the least affected by amyloid fibrils and plaques compared to the rest of the brain – at least in the beginning of the disease. This caused much head-scratching. If beta-amyloid really was the cause of Alzheimer's disease, should it not accumulate mainly in the medial temporal lobe and hippocampus which are the first regions to show nerve cell death and therefore memory symptoms in people with Alzheimer's disease?

Instead, other regions of the brain – such as the frontal cortex and medial parietal cortex – are more seriously affected by beta-amyloid fibrils and plaques – at least in the beginning stages of the disease. Even more confusingly, the regions affected by beta-amyloid don't show the typical nerve cell death we would expect if beta-amyloid alone caused Alzheimer's disease. Understandably, these anatomical findings led many scientists to reconsider whether beta-amyloid was the sole cause for Alzheimer's disease.

When it became possible to measure beta-amyloid levels in living people it was soon discovered that some older people with quite significant amounts of beta-amyloid in their brain lived healthily with no symptoms of Alzheimer's disease. At first, it was thought that these people might eventually develop Alzheimer's disease and were just very resilient to high levels of beta-amyloid. However, it soon turned out that some of those people never developed Alzheimer's disease. This clearly contradicted the original Amyloid Cascade Hypothesis.

The final straw for the original Amyloid Cascade Hypothesis came in a drug trial in the 1990s when a pharmaceutical company developed an Alzheimer's disease vaccine (we'll delve into how this works later on). The vaccine was meant to help the body remove the beta-amyloid more efficiently from the brain and prevent people from developing Alzheimer's disease. The first results looked promising in that the vaccine worked and the body removed more beta-amyloid from the brain. However, to everyone's surprise, some of the trial participants who had reduced amounts of beta-amyloid in their brains still developed Alzheimer's disease. According to the original Amyloid Cascade Hypothesis, this should not have happened since beta-amyloid alone was supposed to cause Alzheimer's disease. So how did the participants in this relatively successful trial still come to develop the disease?

For another group of dementia scientists, these findings were not so surprising. For many decades prior to this, they had been investigating another protein which had been implicated in Alzheimer's disease and had also already been described by Alois Alzheimer and Oskar Fischer – Tau.

23.
TAU

Let's remember that Alois Alzheimer (and Oskar Fischer) not only reported millet sized-seeds outside of nerve cells. He also reported nerve cells which had 'disintegrated', leaving nerve cell fibres [fibrils] behind. Now, we need to be careful here to distinguish amyloid fibrils from tau fibrils. We already know that amyloid fibrils accumulate outside the nerve cell when there is an excess of beta-amyloid. Tau fibrils are different. They are, as Alzheimer's correctly observed, the remains of nerve cells which have died. In summary, beta-amyloid may be found outside of nerve cells but tau is found inside nerve cells or their remains.

Let's first have a look at the role of tau in healthy nerve cells before we can explore how tau contributes to Alzheimer's disease.

Tau is a common protein in the brain. Its main function is to stabilise and drive the assembly of microtubules in nerve cells. Microtubules are, as the name implies, tiny tubes found in large numbers in the nerve cells, where they have two primary roles. Their first role is to provide the nerve cell with structure. Since microtubules are quite rigid, they provide a sort of backbone to the nerve cell. Indeed, they are often counted as part of the cytoskeleton – the skeleton of the cell (cyto from Greek 'cell').

The second main function of microtubules is to transport 'stuff' inside the nerve cells. And with stuff, I mean stuff. Microtubules are like haulage companies, transporting pretty much anything the nerve cell needs to be moved from one place to another. This includes other proteins, nutrients or any other molecules. Things to be moved are usually packed into transport boxes called vesicles. These vesicles can contain almost anything the nerve cell might need at another place. But microtubules do not stop there, they can even transport whole organs of the cells (so-called organelles), like mitochondria and even genetic material, like chromosomes. In a nutshell, if anything needs to be moved, microtubules will move it.

Microtubules can be found anywhere in the nerve cells but there is a concentration of them in a structure called the 'axon', which nerve cells use to communicate with other nerve cells. A nerve cell can have one or more axons and they are vital for nerve cells to communicate with each other. Ultimately, many thousands or even millions of nerve cells communicating together provide our healthy brain function. Axons therefore play a crucial role in our brain's function, and their failure can seriously damage our brain.

This is all very interesting, but what does this have to do with tau and Alzheimer's disease?
Microtubules need tau to work correctly. Indeed, tau belongs to a class of proteins called microtubule-associated proteins (MAPT). How does tau contribute to microtubules working properly?

To understand this we have to go back to how the microtubules transport 'stuff'. Since microtubules are tiny tubes, we would assume that the 'stuff' to be transported will go through the tubes. That would make perfect sense but is in fact wrong. The main reason why the material needing to be transported does not go through the tubes is that the tubes are tiny and so it would limit the size of what can be transported by the diameter of the tubes. Whole cell organelles are many times bigger than the microtubules, so would not move through their tiny tubes. Instead, nature has devised an ingenious way of transporting any shape and size by making microtubules similar to conveyor belts. Here, we need to challenge our imagination a little bit, as there is no easy way – not one that I have found yet at least – to describe how this conveyor belt works. The reason why it is challenging to imagine this 'microtubule conveyor belt' is that the actual conveyor belt is not static, like a conventional conveyor belt. Instead, it moves.

A static conveyor belt would simply stay in one place and move things along its length. But it would be very time-consuming and inefficient for the nerve cells to build long, static conveyor belts. And a static conveyor belt would not work very well. A static conveyor belt would not be able to pick up things from where they need to be moved to where they need to go. Instead, the nerve cells would require another transport mechanism to get things to the conveyor belt. So nature has constructed a more flexible, smaller conveyor belt which can move between any places within the nerve cell.

How does this moving conveyor belt function?
This is where our imagination comes in, as we need to visualise a scenario in which the microtubule conveyor belt assembles and disassembles itself all the time. In biological terms, this is called a dynamic instability system, as the whole system constantly assembles and disassembles itself in order to move along.

Sounds incredible, doesn't it?

But if we extend this analogy of the conveyor belt, we get a fuller picture of what's going on. So, let's say we were to constantly build new conveyor belt parts at one side of our belt, then it would 'move' in that direction. At the same time, if we disassembled part of the conveyor belt at the opposite side, it would then steer it away from this direction. This is how the microtubule system works by constantly adding new parts to the conveyor belt according to where it wants to go and removing parts from the direction it has come from.

Now, the microtubule-associated proteins, such as tau (there are others), are key for the assembly and disassembly of this magical moving microtubule conveyor belt as they link the components of the microtubules together. When the microtubule assembles, the microtubule-associated proteins such as tau link or staple the new conveyor parts together and at that particular disassembly stage, the microtubule-associated proteins simply disconnect from the microtubule so that conveyor parts can be used again. So they help create and then hold in place the conveyor belt.

Microtubule-associated proteins have one other function. If we look at the side of the conveyor belt, we can see that the protein has attachments that stick out from the belt and these attachments allow the vesicles (transport boxes) to be fixed in place and moved along its length.

In summary, then, the microtubule-associated proteins have three key functions: 1) They help assemble and disassemble the microtubules conveyor belt. 2) They keep the conveyor belt stable and connected. 3) They help transport items along the conveyor belt.

Let's now explore what specifically happens to tau in Alzheimer's disease. There is one tau-related process which is particularly relevant to the development of Alzheimer's disease. It is called the hyperphosphorylation of tau. Hyperphosphorylation is quite a mouthful, but we can digest its meaning without too much trouble.

24.
'WHO ORDERED ALL THE PHOSPHATE?'

Hyperphosphorylation of tau simply means that excessive (hyper) phosphate attaches to the tau molecules. In general, tau needs some phosphate to work correctly. But when there is excessive phosphate, tau no longer works properly.

Where is this excessive phosphate coming from and why is there too much phosphate attaching to tau in the first place?
The exact mechanisms of why excessive phosphate sticks to tau is still not completely understood, so for now we just have to accept that there is too much phosphate and this disrupts the normal functioning of tau.

So, what effect does the excessive phosphate have on tau's function?
We already know that tau has three main functions in helping the microtubules work correctly (assembly/disassembly, structure, transport). Excessive phosphate disrupts all these tau functions to some degree, but none more so than the assembly of the microtubule conveyor belt. The reason is that the excessive phosphate 'blocks' all the attachment points that tau needs to affix onto the microtubules.

When this happens tau can't link or staple the different microtubule conveyor belt parts together with each other. Even if some tau sticks to the microtubule, the structure is wobbly since not all of the tau links or staples are present. The microtubule conveyor belt is no longer fit for purpose. The other effect of this reduced linkage of microtubule conveyor belt parts is that it slows down the movement of the microtubule. Remember that the microtubule needs the constant assembly and disassembly of its parts to move along. If, however, this process gets disrupted then the microtubule fails to move along at all or moves very slowly. This means that the transport of vital material that the nerve cell might need is compromised. The net result is that much less vital 'stuff' is transported along the microtubules.

If that weren't already bad enough, the excessive phosphate also makes the tau more 'sticky'. As with beta-amyloid, tau molecules start forming fibrils or oligomers by glueing themselves together lengthwise. This means that even less tau can be now used as it is all gummed up together and cannot perform its three jobs on the conveyor belt. Now, if this process were to happen to one or even a few microtubules, then the nerve cell would be able to cope with it, as there are plenty of other microtubules available. However, we have to imagine that this hyperphosphorylation happens on a much larger scale, affecting, in the end, nearly the whole nerve cell. This means that a significant proportion of the nerve cells transport mechanism is damaged.

For a moment let's take this idea out of the body and demonstrate what it would mean in society. Imagine if the whole haulage industry were to go on strike. The impact on a country would be devastating. It is the same for the transport system in the nerve cell as there would be no – or very little – movement of critical nutrients or cell parts. Some parts of the nerve cell would be starved of nutrients, while other parts might not get the material they need to carry out repair work. The whole nerve cell system starts to collapse. This collapse is particularly prevalent in the axon of the nerve cell, because it has a high concentration of microtubules and tau. Axons help nerve cells to communicate with each other. So the collapse of the microtubule system in the axon has a direct effect on the communication between nerve cells, which is vital for our brain functioning (see also Figure for how this process causes the microtubules to be affected).

Once the nerve cell has collapsed and died, it disintegrates and most of its remaining debris is cleaned away or recycled by our body but just like the beta-amyloid fibrils tau fibrils are a sticky, difficult mess for the body to remove or recycle and are often the only thing left over from the collapsed and disintegrated nerve cells.

Now we know the basic physiological processes of how tau causes nerve cell death. However, there is still one mystery. If only a few nerve cells are affected by tau hyperphosphorylation, why does the disease spread through the brain, causing an increasingly large area of the brain to be affected and eventually leading to Alzheimer's disease? Surely, the brain could cope with losing some nerve cells without developing Alzheimer's disease. Why is tau so important in the development of Alzheimer's disease?

To try and answer some of these questions, we have to understand one final – still controversial – aspect of hyperphosphorylated tau, and that is its remarkable ability to 'infect' other nerve cells.

Right: ALOIS ALZHEIMER, the German doctor after whom Alzheimer's disease is named.
Fig 1: Creative Commons, TC Üsküdar University

Above: THE HOSPITAL FOR THE MENTALLY ILL AND EPILEPTICS near Frankfurt, where Alzheimer met Auguste Deter.
Fig 2: Creative Commons, Architecture Museum TU Berlin

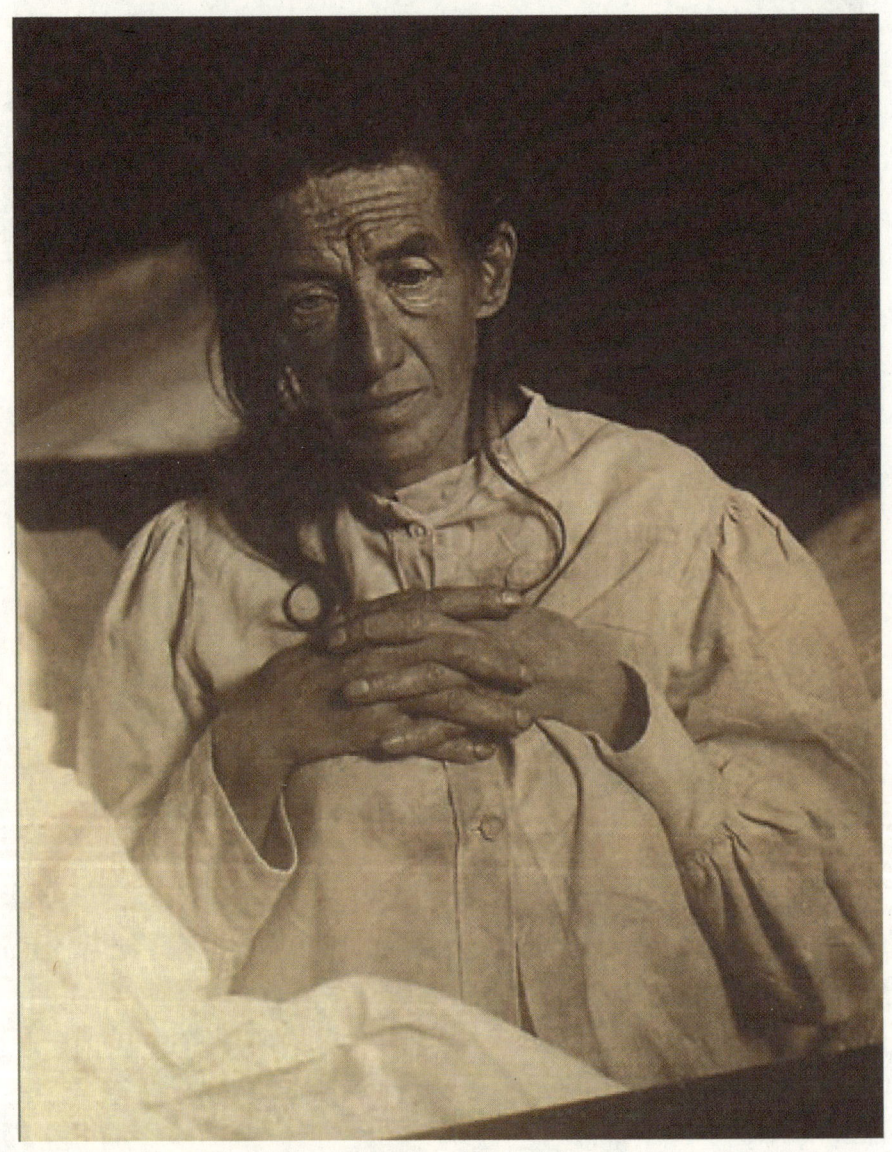

Above: AUGUSTE DETER, woman in her fifties who showed extensive memory loss when interviewed by Alzheimer. Fig 3: Creative Commons, public domain.

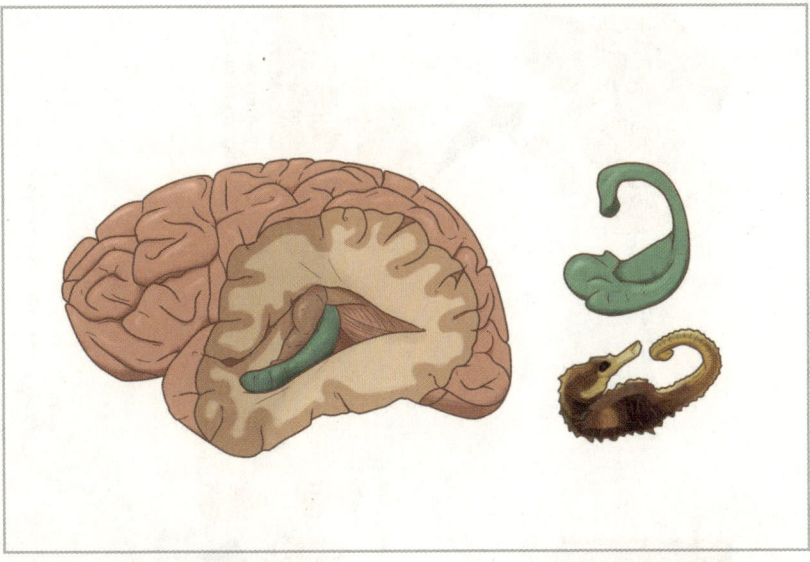

Above: THE HIPPOCAMPUS (in green), the area of the brain where memories are indexed and which affected by Alzheimer's disease early on. It is named after the Greek word for seahorse, whose shape it resembles. Fig 4: Michael Hornberger,

Above: OSKAR FISCHER, a pioneering Jewish psychiatrist. Fischer described similar cases to Alzheimer but his name has been largely forgotten. Fig 5: Creative Commons, public domain.

Right: FISCHER died while being held in the concentration camp of Theresienstadt
Fig 6: Creative Commons, Daniel Baránek and Emmanuel Dyan.

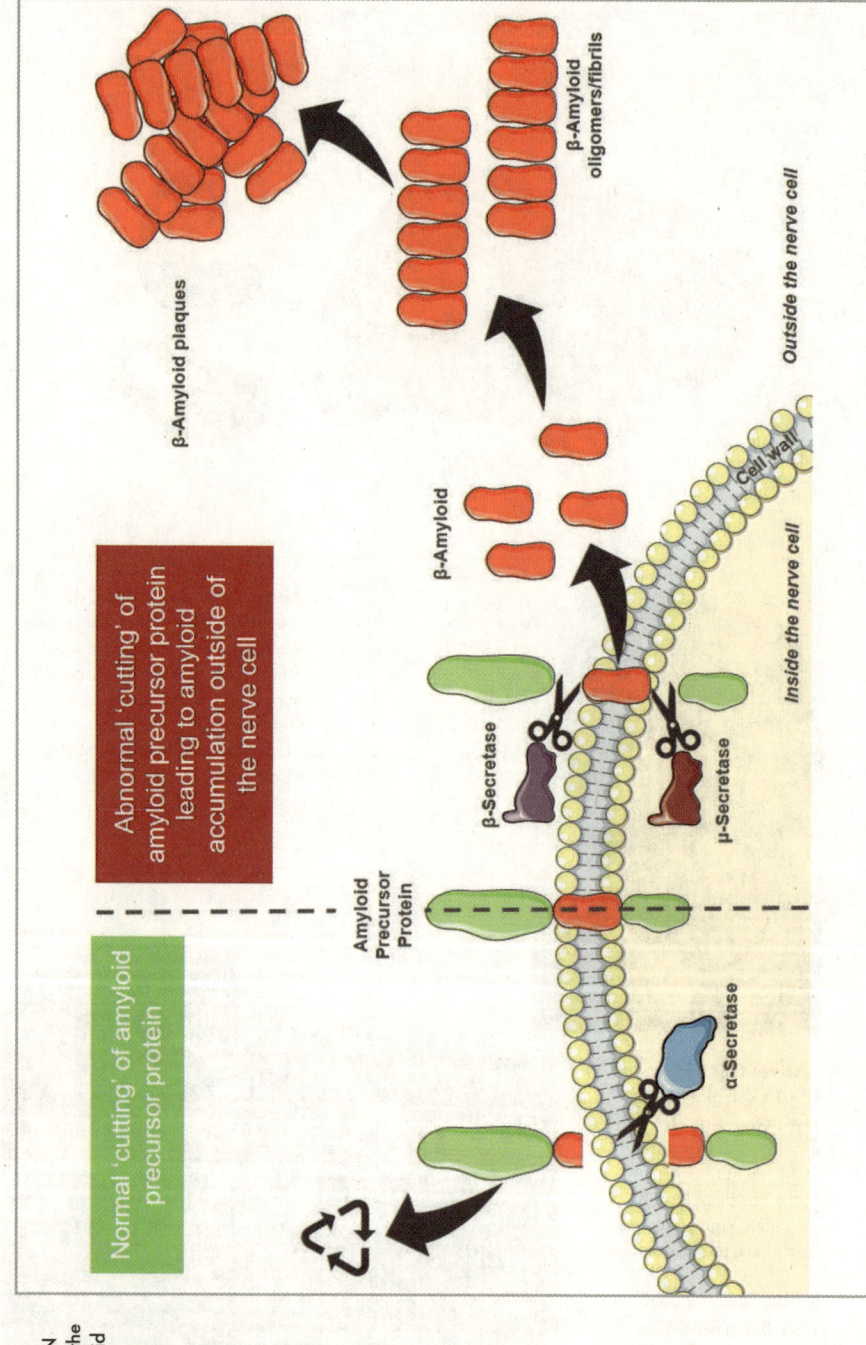

Right: ILLUSTRATION showing how the protein amyloid accumulates in Alzheimer's disease. Fig 7: Michael Hornberger.

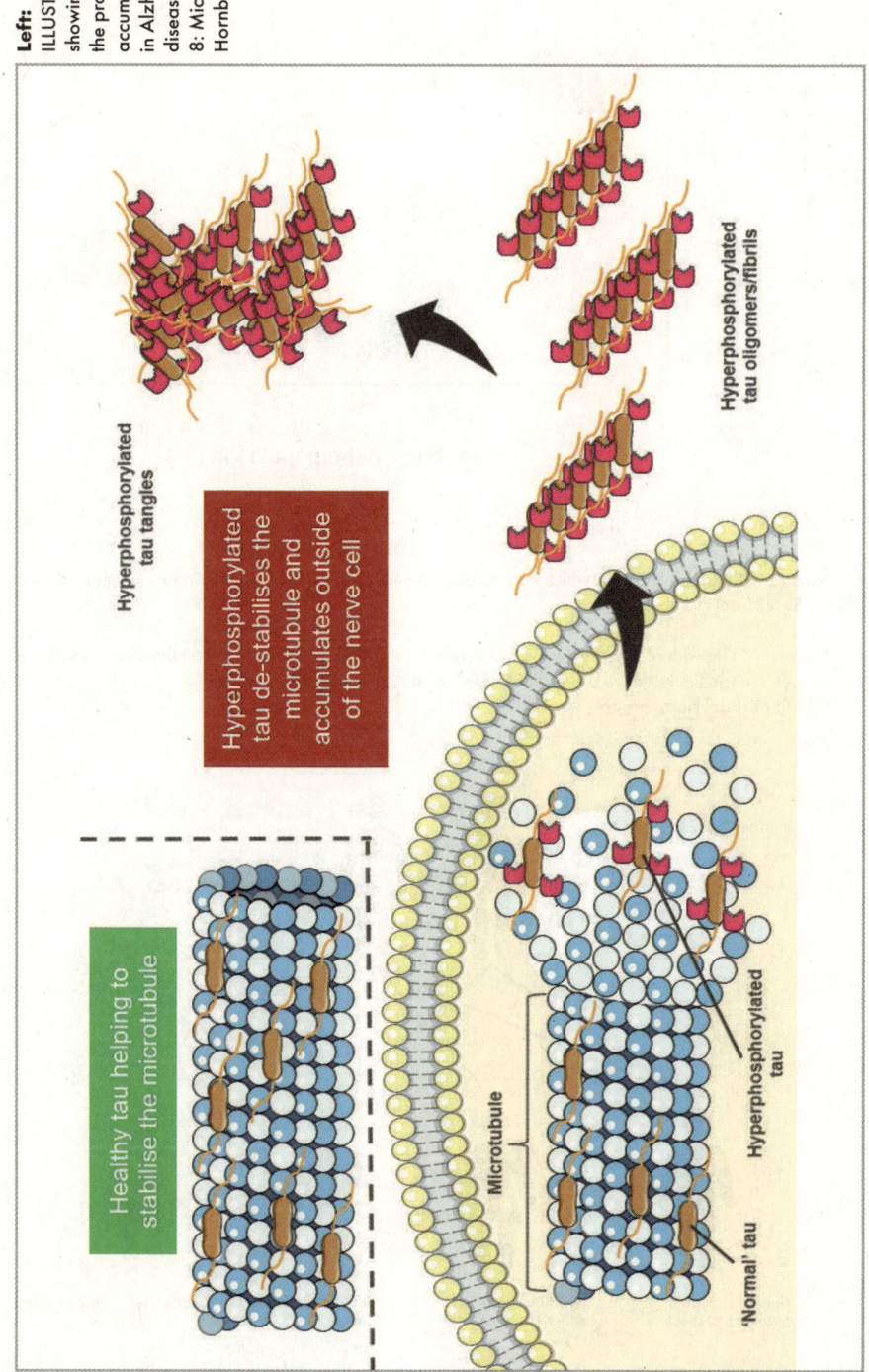

Left: ILLUSTRATION showing how the protein tau accumulates in Alzheimer's disease. Fig 8: Michael Hornberger.

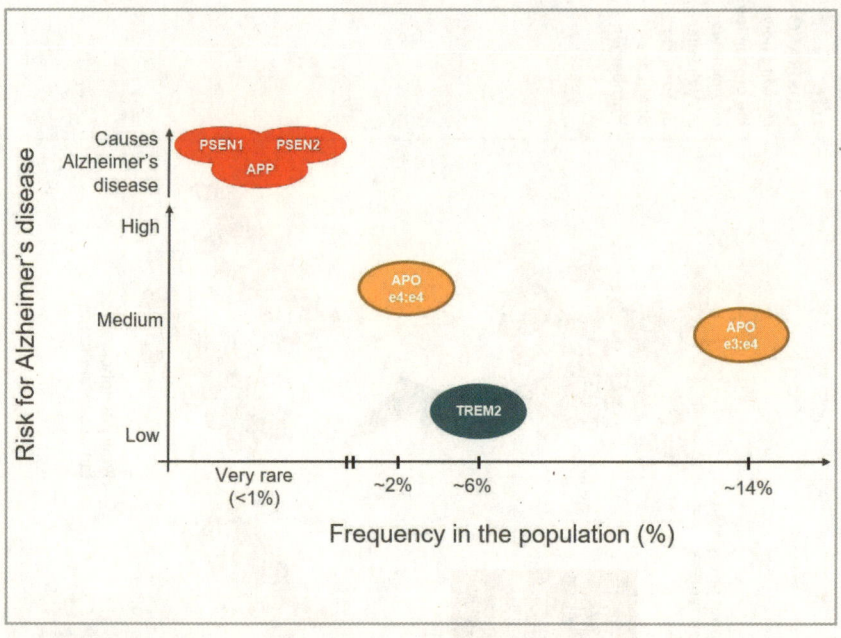

Above: GENETIC variations which can increase or reduce our risk of developing Alzheimer's disease.
Fig 9: Michael Hornberger.

Below: AN individual inherits a gene from each parent. When certain gene combinations are passed on (see orange figure below), the risk of developing Alzheimer's disease rises.
Fig 10: Michael Hornberger.

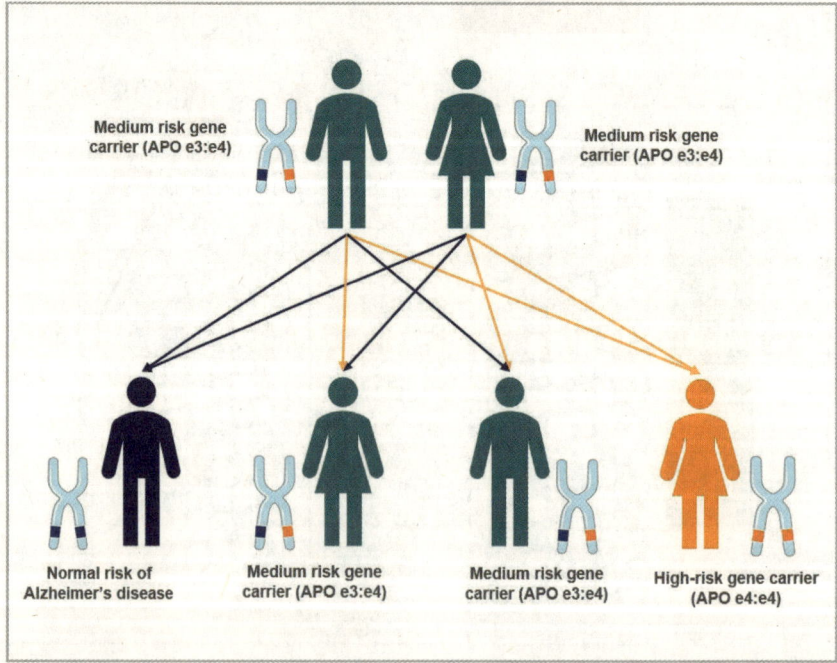

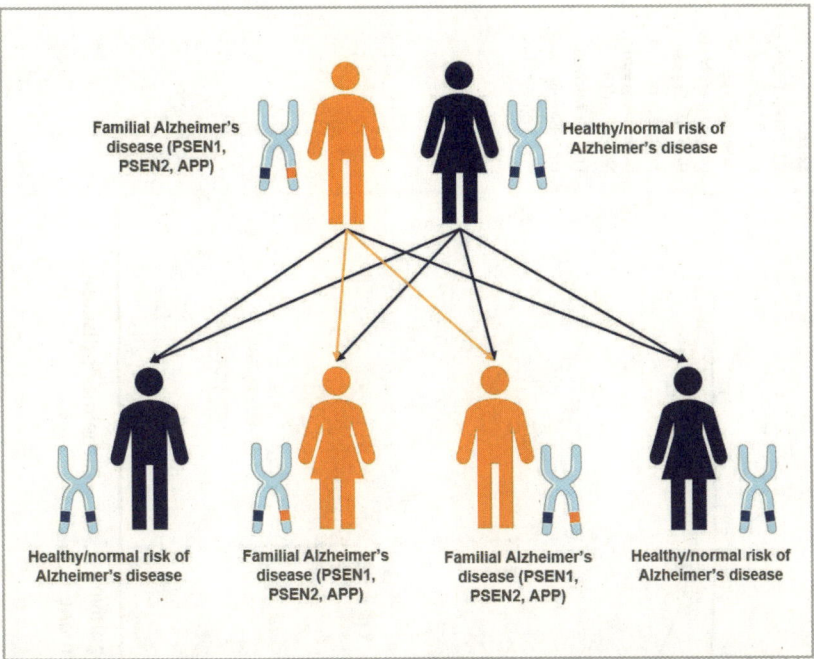

Above: FAMILIAL Alzheimer's disease is very rare and passed down through families, with children having a 50% chance of getting the gene which causes Alzheimer's. Fig 11: Michael Hornberger.

Below: THE APOE gene is the most common risk factor for Alzheimer's disease in the population. Around 78% of the population has the e3:e3 APOE genotype, which does not increase our risk for Alzheimer's disease. Other APOE genotypes can increase or decrease our risk for Alzheimer's disease. Fig 12: Michael Hornberger.

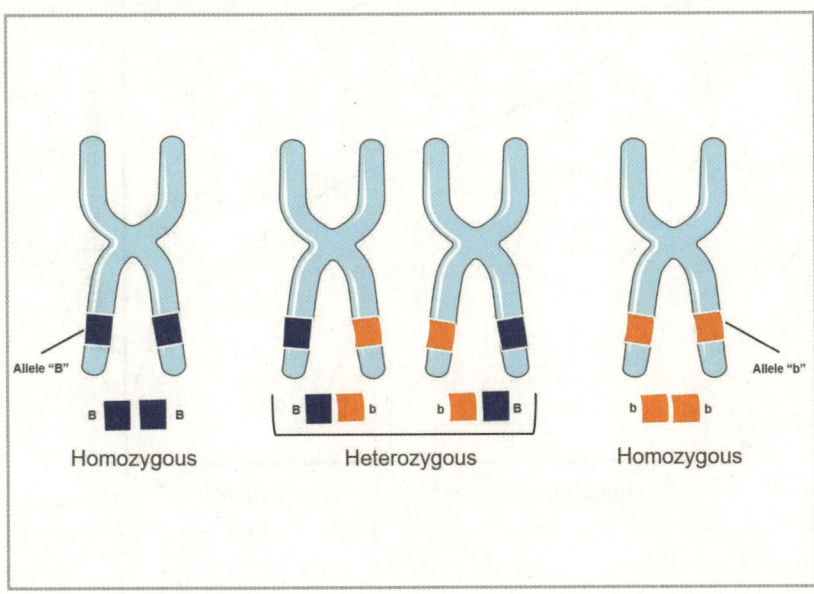

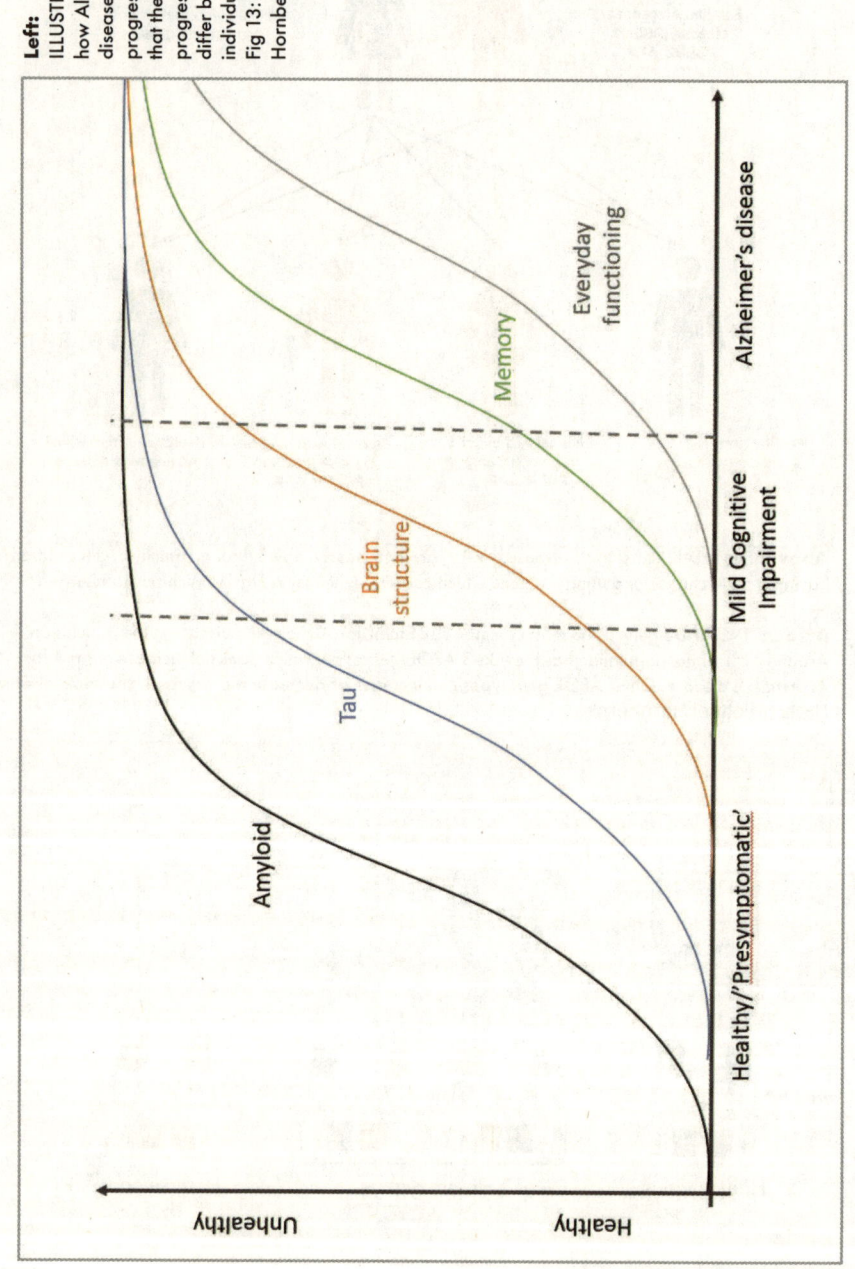

Left: ILLUSTRATION of how Alzheimer's disease typically progresses. Note that the disease progression can differ between individuals.
Fig 13: Michael Hornberger.

25.
TAU 'INFECTION'

We usually think of an infection as an immune system reaction to an external pathogen threatening our body. For example, a virus or bacterium, which causes damage to our body, like the SARS-CoV-2 virus which causes Covid-19. Our body's immune system tries to fight off this 'infective invasion' by a virus or bacterium. This is pretty much a correct assumption for nearly all infection processes.

But there are also rarer types of infection, which do not need bacteria or viruses. Instead, these rarer infections occur because of wrongly built proteins. If a wrongly built protein causes an infection, it is called a 'prion'. Some readers in the UK may recall prions from the time when there was a large outbreak of the so-called 'Mad Cow Disease,' or bovine spongiform encephalopathy (literally: porous brain damage in cows), which is caused by this kind of wrongly built protein. In humans, the most common form of prion disease is called Creutzfeldt-Jakob disease, which is extremely rare but can be fatal.

Before people start to panic that Alzheimer's disease might also be infectious like Mad Cow Disease, do not worry. It is not possible to get a human prion disease from ordinary day-to-day contact, or by airborne droplets, blood or sexual contact, which are the most common forms of infectious transmission. Research has shown that even people working with brain tissue of people with Alzheimer's disease, such as neuropathologists or neurosurgeons, are not at greater risk of being 'infected' with Alzheimer's disease.

What is the evidence then that Alzheimer's disease is potentially a prion disease?
There is in fact no hard evidence yet. However, laboratory experiments have shown that tau phosphorylation seems to spread to neighbouring nerve cells, which in turn die. At this stage we do not completely understand how tau hyperphosphorylation can 'infect' neighbouring nerve cells. In light of this, some scientists have proposed the idea that tau hyperphosphorylation has 'prion-like' characteristics. Basically, hyperphosphorylated tau seems to behave like a prion but we still don't know whether tau is actually a prion and even whether it actually causes the onset of Alzheimer's disease.

It would make sense if hyperphosphorylated tau did cause Alzheimer's. We can track the nerve cell damage caused by tau phosphorylation along the nerve cells which are connected and communicate with each other. It 'infects' more and more nerve cells which are all connected to each other, until most of the brain is affected by the tau hyperphosphorylation and eventually nerve cells die. Again, Alois Alzheimer was perceptive enough to see this and described how, in Auguste Deter's brain, 'between one-quarter to three-quarter of nerve cells' were affected by this 'unknown substance'. This substance we now know is hyperphosphorylated tau. Tau phosphorylation affected nearly three-quarters of Mrs Deter's brain at her death, even though it must have started in just one nerve cell. And we know that the higher the levels of phosphorylated tau in an individual's brain, the higher the likelihood that person has Alzheimer's disease.

That's all very well but I'm confused now. Does this suggest that it's actually tau hyperphosphorylation and not beta-amyloid which causes Alzheimer's disease?
Both hyperphosphorylated tau and beta-amyloid have a strong association with Alzheimer's disease. Indeed the current scientific

consensus is that for most people, we need both – beta-amyloid and phosphorylated tau – in our brains to develop Alzheimer's disease.

I wish we could leave it at that, but for the sake of completeness, I need to mention that even though the co-occurrence of higher beta-amyloid and higher phosphorylated tau explains (or seems to explain) the development of Alzheimer's disease in most people, it does not for others. A small minority of people can develop Alzheimer's disease even without any excessive beta-amyloid in their brain.

What a mess, you may be thinking.

You would be right. But it is important to give you a full idea of this 'mess' as it helps to explain why the scientific community has been confounded for decades as to 'why' Alzheimer's disease develops. We do, however, have a good knowledge as to 'how' Alzheimer's disease develops.

26.
THE 'EMERGENCE' OF THE DISEASE

Now that we have come to mention the 'how' of Alzheimer's disease, what is the latest scientific model of how the disorder emerges?

Scientists have since come up with a more comprehensive model for how Alzheimer's disease develops. The new model still features a cascade of steps which must occur, but it now incorporates a timeline to indicate 'when' those changes occur.

Below, I have included a graph of the accepted scientific model of the emergence of Alzheimer's disease to better illustrate the cascade of the disease-specific processes which will ultimately lead to memory-related symptoms and everyday function changes. However, please note that this is just the latest available model – and existing data both confirms and refutes this model for Alzheimer's disease. As with any model it is useful but not necessarily correct. Finally, scientific models show the 'ideal' scenario shown by the data, so we need to keep in mind that other external factors, such as lifestyle and the environment, might influence the trajectories of the model for each individual. But as a general model of Alzheimer's disease, this is an excellent way to track how the disease will emerge in most people.

Let's go through this graph. We can see the x-axis (horizontal line) shows the progression of the disease from healthy/'presymptomatic' all the way to Alzheimer's disease. With the term presymptomatic we mean people who have the initial changes associated with Alzheimer's disease in their brain but have not yet developed symptoms, hence they are referred to as presymptomatic (pre from the Latin preposition 'prae' = 'before'). The y-axis (the vertical line) shows us how the level of disease changes, with lower values showing 'healthy' levels and higher levels showing 'unhealthy' levels. The exact threshold where healthy changes to unhealthy for these variables is still a matter for debate and, for now, we just have to accept that at some point the factors change from healthy to unhealthy. Finally, the dashed lines show potential transition points from Healthy/'Presymptomatic' to Mild Cognitive Impairment to Alzheimer's disease. The dashed lines are merely to give an indication when someone is likely to receive a diagnosis of Mild Cognitive Impairment or Alzheimer's disease. Again, this might differ between individuals, as some might report their symptoms earlier or later, and similarly, doctors may diagnose someone earlier or later.

We are ready now to explore this graph in more detail.

In the graph, we can see five lines, each one a key factor in the development of Alzheimer's disease and its eventual diagnosis. Two of the lines (black and blue) refer to the proteins accumulating in the Alzheimer's disease process, amyloid and tau (the higher the line, the higher the levels of amyloid and tau). The orange line refers to the brain structure, and it measures how much the brain cells are affected by the disease process caused by the proteins. We know already that the more amyloid and tau accumulate the more the brain and particular nerve cells become affected since a large concentration of proteins are toxic to the nerve cells and these cells eventually die. If this nerve cell death happens on a large scale we see changes to the brain structure, as the large nerve cell loss causes the brain to shrink (the brain shrinkage is called atrophy by clinicians – deriving from the Ancient Greek word 'atrophia' = 'wasting away'). The changes to brain structure covered in the graph therefore denote an increased level of nerve cell death which we will see as atrophy on brain imaging scans. Next, the green line tracks the typical memory symptoms we see in Alzheimer's disease when people have problems recalling recent events or can't remember where they placed things. The higher the line the more memory problems there are. Finally, the grey line represents how the everyday functioning of people is affected by the disease, again the higher the line the more everyday functioning is affected.

Now that we know what the lines represent, we can see that there is a cascade of events happening before people develop memory symptoms and receive a diagnosis. First off, it is amyloid which increases in people's brains and this process is thought to take anything from a year to several decades to occur, so the slope of the curve in the graph can clearly change. We can also see that amyloid levels are already very high before the next protein (tau) starts accumulating significantly in the brain. Several theories posit that high levels of amyloid are required for tau to accumulate in Alzheimer's disease. However, the interaction of amyloid and tau is complicated and inconclusive.

The next line (orange – brain structure changes) starts only once both proteins (amyloid and tau) are widely distributed in the brain. This makes sense since significant amounts of amyloid and tau need to have accumulated in order to become toxic to the nerve cells. Once the cells start dying this then causes the atrophy that is visible on brain scans. Shortly after the brain changes emerge, memory problems become noticeable. Again, this makes sense since the intact nerve cells are required for our memory to function normally and once a substantial quantity of nerve cells starts dying our memory can no longer function and our forgetfulness becomes a problem. It is usually at this stage that we report our memory symptoms to our doctor who will give us a diagnosis of Mild Cognitive Impairment if we meet all the diagnostic criteria.

Finally, once the memory symptoms start to worsen, it will affect our everyday functioning and other cognitive symptoms might also emerge, which will then lead to a diagnosis of Alzheimer's disease by a doctor.

The model should make clear that there is a defined cascade for most people who develop Alzheimer's disease. But instead of relying 'simply' on the Amyloid Cascade it is the interaction of amyloid and tau which is the critical factor in the emergence of the disease. Scientific data supports this model to a large degree by showing that protein levels of amyloid and tau rise first before we see brain changes and the subsequent memory and everyday changes. It also means that the typical memory symptoms, which lead to a diagnosis of Alzheimer's disease, only emerge quite late in the disease process. This is sobering news for many people since the emergence of the memory symptoms means that the disease has already gripped the brain. It also explains why at this stage many treatments fail to have much impact, as the damage to the brain has already been done.

Why not detect people then in the earlier/presymptomatic stages of the disease, so that we can treat Alzheimer's disease earlier and better?
It seems pretty obvious that if we want to treat the disease better we should detect it much earlier, ideally when amyloid and tau are still accumulating and have not badly damaged nerve cells. Until very recently, we did not have the technology to measure amyloid and tau in living people. Thankfully, now we can.

27.
BIOMARKER TECHNOLOGY

Technological advances have always spurred scientific discovery. Such new technologies usually allow us to measure natural processes that we had been previously unable to observe. Remember, for example, the emergence of microscopy that allowed Alois Alzheimer and Oskar Fischer to make their discoveries. Up until a few years ago it was quite difficult to measure amyloid and tau levels in living people. Most studies investigating the cause of the disease were conducted on people who had died from Alzheimer's disease, as with the post-mortem studies conducted by Alzheimer himself at the start of the 20th Century. In particular, during the 1980s several post-mortem studies showed how amyloid and tau progress through the brain. The studies did so by comparing the cases of people with Alzheimer's disease who had died at different stages of the disease. These studies boosted our understanding of the genesis and propagation of the disease, but we could not follow the onset and progression of Alzheimer's in living people until the development of Alzheimer biomarkers in the 1990s.

What are biomarkers?
Simply put, biomarkers are biological markers of a disease. For Alzheimer's disease, biomarkers enable the detection of beta-amyloid and phosphorylated tau in living people. A key discovery in

the development of Alzheimer's disease biomarkers in living people was the realisation that beta-amyloid and phosphorylated tau might not only stay in the brain but also get secreted into the blood and the corticospinal fluid around our brain and spinal cord (protecting and nourishing them and flushing away waste products, such as amyloid and tau.) New molecular techniques, such as the ELISA (Enzyme-Linked Immuno Sorbent Assay) technique, reliably measure these low levels of beta-amyloid and phosphorylated tau. This means we can measure amyloid and tau in people with Alzheimer's, even before they develop symptoms and get a diagnosis. As we have already seen, the accumulation of the proteins can start many years, if not decades, before the symptoms start. So, detecting increased levels of beta-amyloid and phosphorylated tau in people before they develop symptoms – so-called presymptomatic Alzheimer's disease – is key to the early prevention or treatment of the disease.

If these techniques exist, why are they not commonly offered as part of the dementia diagnostics?
The most obvious reason why corticospinal tests are not being routinely offered for dementia diagnostics is that they require a lumbar puncture. A lumbar puncture is a medical procedure, during which a long needle is inserted between the vertebrae of people's back to collect a very small amount of corticospinal fluid. Despite the lumbar puncture being a routine medical procedure these days, it is quite an invasive procedure, and usually requires a short stay in the hospital, with a specialist doctor conducting the procedure. It is therefore not a simple 'test' which can be conducted by our family doctor or nurse, but requires hospital admission. That might be fine if we need to do this once but we might have to repeat lumbar punctures to measure elevated levels of amyloid and tau over time. This is clearly not feasible and hence corticospinal fluid tests for amyloid and tau have been mostly conducted as part of research studies to investigate the development of the disease. For everyday clinical diagnosis it is too risky for everyone to undergo repeated lumbar punctures, not to speak of the costs and infrastructure which would be required.

But why do we need a difficult lumbar puncture to measure beta-amyloid and phosphorylated tau, when we can also measure them in the blood?

That would be much simpler as even our family doctor or a nurse could take a blood sample and send it off for analysis. No need for going to the hospital and having a procedure which is far more risky than blood taking. A blood test for Alzheimer's disease has been the 'holy grail' for many years. But only now blood tests to measure beta-amyloid and phosphorylated tau reliably are emerging. These tests are still in development but look now very promising in detecting amyloid and tau in the presymptomatic stages of Alzheimer's disease. The reason why it took so long to develop the blood tests is that beta-amyloid and phosphorylated tau levels are even smaller in the blood than in the corticospinal fluid. So, it required new technological advances to amplify these levels to measure them reliably.

Blood tests for Alzheimer's disease will clearly transform the diagnosis of people before they develop any symptoms over the coming years. It will allow for the first time to treat or potentially even prevent Alzheimer's disease in people before they develop symptoms. Since we now know that symptoms only occur when the disease has already taken a grip on the brain, the earlier detection of amyloid and tau will provide a much better chance to fight the development and progression of the disease.

However, the amyloid and tau levels in the blood or corticospinal fluid are only an indicator of the amyloid and tau levels in the brain, not a precise measurement. The other issue is that based on the blood tests, we do not know where in the brain the amyloid and tau is accumulating. Brain imaging techniques can actually measure and detect beta-amyloid and phosphorylated tau in the brain. These techniques involve a brain scan called Positron Emission Tomography (PET). For Positron Emission Tomography to work a person gets an injection or drip with a solution which has some radioactive tracers attached to its molecules. Once the solution is in the body, it can detect specific changes in the organs, which can be then measured via the radioactive signal from that organ in the Positron Emission Tomography scanner. One of the most commonly used brain tracers (as those solutions are commonly called for PET) is fluorodeoxyglucose. Fluorodeoxyglucose is sugar (glucose) which contains a modified, radioactive fluoride molecule. Since our cells use glucose for their activity we can measure the levels of cell activity with this tracer. If we are injected with this tracer and are put inside a Positron Emission Tomography machine, clinicians can measure the levels of our brain activity. Less active (hypoactive) areas indicate that the nerve cells have started dying in these regions and therefore tell us how Alzheimer's disease is affecting the brain.

In the 2010s, several research groups started developing Positron Emission Tomography tracers specific for beta-amyloid and phosphorylated tau, meaning we can now measure where and how much amyloid and tau have accumulated in a person's brain. Similar to the beta-amyloid and phosphorylated tau blood tests, the Positron Emission Tomography scans for beta-amyloid and phosphorylated tau are still experimental and mostly used for research studies and not clinical service. Despite their promise Positron Emission Tomography scans are unlikely to be used routinely for dementia diagnostics. The reason is that such scans can only be conducted at specialised centres and are quite expensive. Still, for unusual or rare forms of dementia Positron Emission Tomography scans might provide an alternative to blood tests to identify the underlying dementia.

Technological biomarker advances have evidently made a huge difference to the detection of beta-amyloid and phosphorylated tau and will improve future diagnosis of people with Alzheimer's disease or those at risk of Alzheimer's disease. So what? You might say. If there is still no treatment for the disease available, what is the point of detecting abnormal beta-amyloid and phosphorylated tau levels in people before they develop the disease? Is this not a bit cruel, if not unethical, to tell people that they are likely to develop Alzheimer's disease without being able to offer them some treatment?

These are all fair questions. However, we should not underestimate how existing prevention techniques and medication can alter our risk of developing Alzheimer's disease. We will discuss this in more detail in Part IV of the book when looking at risk factors. Suffice to say that the current scientific consensus is that existing prevention techniques and medications can change our risk of developing Alzheimer's disease by up to 40%, which is an astonishingly large figure. However, the key for successful risk reduction or prevention is to identify the onset of the disease as early as possible, when these techniques and medications are most powerful. If we know that we are at increased risk of Alzheimer's disease, the new biomarkers will make it more likely that we can at least slow down the development of the disease with lifestyle improvements and medication. We can also now measure in a living person if a new Alzheimer's disease treatment reduces beta-amyloid and/or phosphorylated tau levels, and in turn whether it slows down or even stops the disease.

Let's now have a look at the treatment approaches for Alzheimer's disease to understand how they work and which ones look the most promising.

28.
AMYLOID TREATMENT APPROACHES

Over the past few decades dozens of studies have trialled new drugs and treatments for Alzheimer's. While they have failed to find a cure, we do now have a range of new medications available which can slow down the disease. To better understand why so many drug trials have failed and why we cannot yet stop Alzheimer's disease, we will have a look now at the pharmacological treatment approaches targeting amyloid and tau. Given that the Amyloid Cascade Hypothesis held sway for so many years, it is not surprising that most clinical trials for new Alzheimer's disease treatments focused on the accumulation of beta-amyloid.

Drug treatment approaches for beta-amyloid can be divided into three main categories:

1. Decrease of beta-amyloid production.

2. Prevention of beta-amyloid accumulation.

3. Increase of beta-amyloid removal.

Let's go through these various types and see how they work and which ones have been shown to be the most successful. It's worth noting that none of these different approaches to the problem is better or worse than the others – they are simply different ways for scientists to change the levels of beta-amyloid in the brain.

Decreasing beta-amyloid production

This one seems almost too obvious. Just decrease the brain's overall production of beta-amyloid and (voila!) we lower the risk of Alzheimer's disease.

But how can we reduce beta-amyloid production in the brain?

The most common approach for these treatment studies is targeting the beta-amyloid secretases. Remember the secretases, which 'snip' amyloid into its different components? The beta and gamma secretases are good treatment targets, as they regulate the production of beta-amyloid. The logic behind this is that the fewer the secretases cut amyloid into beta-amyloid, the less beta-amyloid will accumulate. Specifically, the drugs target the 'behaviour' of the secretases.

One strategy is a drug to reduce the overall activity of the beta and gamma secretases by making the beta and gamma secretases less efficient, for example by slowing them down. Once the beta and gamma secretases are 'inhibited' in their function, they will produce less beta-amyloid, which should result in less amyloid accumulation and plaques, and eventually reduce the risk of Alzheimer's disease. The second secretase 'behaviour' strategy is to decrease the beta-amyloid production by encouraging the other secretase that snip amyloid (alpha secretase) to become more active. Remember that besides the beta and gamma secretase there is also an alpha secretase which snips the amyloid protein into components that are different from the beta and gamma secretases. Importantly, the end product from the alpha secretase snipping amyloid does not produce beta-amyloid. So, if we encourage the alpha secretase to become more active than the beta and gamma secretases, then amyloid will be still broken down but not into beta-amyloid.

The danger here is that if we change the activity of the secretases we might cause some other, inadvertent effects in the brain not related to beta-amyloid production, which might have serious consequences. Such trade-offs are often a dilemma for disease treatment approaches. We want to prevent or treat the disease without disturbing the healthy biological functions, as this could otherwise lead to potential side effects of the treatment. This is particularly true for medication targeting healthy body functions to prevent disease, such as influencing the secretases for beta-amyloid production. Still, targeting the secretases is a worthwhile approach to preventing Alzheimer's disease but it just needs to be investigated very carefully. The next approach takes a slightly different angle towards prevention by targeting the accumulation of beta-amyloid.

Prevention of beta-amyloid accumulation

The rationale behind this treatment approach is that while our brain generally deals with beta-amyloid quite well, problems occur when it starts clumping together into oligomers and plaques. As we know already, removing those oligomers and plaques is much harder for the brain and increases the risk of developing Alzheimer's disease. Avoiding the build-up of oligomers and plaques is therefore the aim of this treatment approach.

There are two main treatment approaches to reduce or avoid the build-up of beta-amyloid oligomers or plaques. The first tries to boost molecules, so-called anti-amyloid aggregators, which are known to reduce the build of beta-amyloid oligomers and plaques. These molecules disrupt the building and 'glueing' together of the beta-amyloid sheets. Remember how we compared this process to sticking lasagne sheets together lengthwise? The treatment molecules basically reduce the sticking together of the beta-amyloid molecule sheets and therefore fewer or even no beta-amyloid oligomers form. The beta-amyloid remains therefore in a 'looser' composition. This makes it easier for the brain to remove it, instead of having to deal with the more effortful removal of beta-amyloid oligomers or even plaques.

Many of those anti-amyloid aggregator molecules are actually part of our diet – or should be. For example, curcumin, a molecule which is part of turmeric, is effective in reducing beta-amyloid oligomer formation. Similarly, a class of molecules called flavonoids, which are common in dark berries and red wine, have similar functions among other benefits. But before you rush to the kitchen to gulp down a spoon of turmeric and a glass of red wine, consider this. Although these molecules are common in these foods it does not necessarily mean that eating a lot of them will stop the accumulation of amyloid. How food is processed by our body is a complex process and it is currently not clear how much of these molecules benefit us or are simply excreted.

Still, for centuries existing foods and plants have helped us to develop new medicine. Medication is often nothing else than a highly concentrated form of a natural molecule in a food or plant. It makes sense that similar molecules might help develop future medication for Alzheimer's disease, but they clearly need to be higher concentrated and targeted towards beta-amyloid. So, hold off on that spoon of turmeric and matching wine course for now.

In any case, the two previously discussed approaches to beta-amyloid are currently dwarfed in terms of their effectiveness by the third one, which has proven to be the most successful one to date.

Increase of beta-amyloid removal

The third treatment approach is to increase the removal of beta-amyloid. For this treatment approach, we are not concerned about reducing the production or build-up of beta-amyloid, instead we are boosting the removal of beta-amyloid oligomers and plaques. With this approach we have clearly moved from prevention (trying to reduce beta-amyloid production or accumulation) to targeting the beta-amyloid itself.

How do we increase beta-amyloid removal?

The best way to increase beta-amyloid removal is to boost our body's own disease-fighting system – the immune system – to remove the accumulated beta-amyloid oligomers and plaques.

How can we boost our immune system so that it removes more beta-amyloid?

The key technology is based on immunisation. Immunisation is mainly associated with children, who can be badly harmed by disease such as measles. However, since Covid-19, we are aware that immunisation can also be vital for fighting potential deadly diseases in adults. Any immunisation therapy works by using antibodies in our body. Antibodies can be regarded as the 'scouts' of the immune system, always patrolling our body on the lookout for potential threats to our body, which need to be dealt with by our immune system. Antibodies are proteins which attach to potential threats, such as unwelcome viruses or bacterias, which our immune system needs to deal with. Each antibody is highly specific to each invading threat (virus, bacterium).

Once the antibody locks onto the bacterium, virus or protein, the antibodies flag it to the immune system to neutralise, removing or destroying the threat. The removal and destruction of threats from the body is carried out by the T-cells and B-cells of the body's immune system. It is a highly efficient system, which has guaranteed our survival throughout evolution. However, the immune system has one weak point. If we do not have the antibody for a threat, our immune system cannot fight that threat efficiently. Again, Covid-19 is a perfect example for this as it was a new virus for which our bodies did not have an antibody. That is why so many people became severely sick or even died of the SARS-Cov2 virus, because our bodies only started to produce antibodies against the virus once we were infected. Some people produced antibodies quickly, resulting in an efficient elimination of the virus by the immune system. However, for others, the antibody production was slower – for unknown reasons – and the virus could take a much stronger hold of the body before the immune system responded. For some people, this delay meant that they got very sick or even died, as the body could no longer deal with the, by then, rampant infection.

Immunisation therapies work on the principle of speeding up this response or providing the body with the antibodies for a particular threat. This means the body can react to the threat of the disease much faster by having the antibodies flagging the threat to the body immediately and the T- and B-cells directly removing the threat.

What has this all to do with beta-amyloid?
Most of us are, of course, familiar with such immunisation strategies when deployed against diseases like measles or Covid-19.

What many people do not know is that we can actually use similar immunisation therapies against the accumulation of proteins such as beta-amyloid. We 'simply' need to create antibodies associated with the beta-amyloid oligomers and plaques, and the body's own immune system will start removing them. It is a brilliant way to use the body's own strength to fight Alzheimer's disease. Immunisation therapies turbocharge the body's immune system, so it can remove the beta-amyloid itself. All the body needs is a helping hand by having the right antibodies. The 'helping hand' for the body to remove beta-amyloid comes in the shape of monoclonal antibodies.

What are monoclonal antibodies?

Monoclonal antibodies are antibodies created in a lab which attach themselves to specific proteins, in our case beta-amyloid. Monoclonal means that the antibodies are all identical. They are literally clones of each other. Once the monoclonal antibodies are injected into the body they will look out for proteins to which they can attach themselves. Amyloid monoclonal antibodies are looking for beta-amyloid. Once they locate the beta-amyloid they attach themselves to the oligomers or plaques and flag it to the immune system. The immune system will notice these flags and move in to destroy these beta-amyloid oligomers and plaques.

To date this method has proven to be the most successful in slowing down Alzheimer's disease. Several monoclonal antibody therapies are now available to people with dementia – a tip to spot them is that most end with the suffix -mab (for monoclonal antibody). At the moment these medications are only licensed for people with mild Alzheimer's disease. Despite their success in removing beta-amyloid from the brain, some of these medications can have significant side effects for some people. These side effects can include fluid or blood pooling in the brain, which only occurs to a very small percentage of people but is very serious. Therefore, people who are prescribed these new medications are currently required to undergo regular MRI brain scans when taking the medication to monitor their brain health. Hopefully, future medications will have milder side effects and less onerous requirements.

Still, these new medications are an enormous step forward, as for the first time they provide a treatment to tackle beta-amyloid. However, we need to make clear that these medications will not be a cure for Alzheimer's disease. Instead, clinical trials have shown that they 'only' slow down the progress of the disease. That makes perfect sense, as we now know that there is still no treatment for phosphorylated tau, which in combination with beta-amyloid causes Alzheimer's disease. By removing beta-amyloid we simply reduce the amount that can react with phosphorylated tau. Let's now focus on the treatment of the second critical protein for Alzheimer's disease: tau.

29.
TAU TREATMENT APPROACHES

Drug development for tau is lagging slightly behind that for amyloid, perhaps because of the initial focus on the Amyloid Cascade Hypothesis. Now, of course we know that it 'takes two to tango' for Alzheimer's disease, namely beta-amyloid and phosphorylated tau. Despite tau treatment approaches having been slightly slower off the blocks, the clinical trials for tau have clearly benefited from the failures of many early amyloid drug developments. Tau treatment approaches can be divided into four categories, namely:

1. Decrease of overall tau.

2. Decrease of tau phosphorylation.

3. Decrease of phosphorylated tau accumulation.

4. Increase of phosphorylated tau removal.

Let's go through each one in turn.

Decrease of overall tau

The decrease of overall tau is a different approach to decreasing beta-amyloid. The goal of this treatment approach is to reduce overall levels of tau, not just phosphorylated tau.

But don't we need tau for our healthy nerve cell function and its role as a microtubule-associated protein?

Yes, we do, but let's remember that tau is only one of the microtubule-associated proteins. Therefore, the nerve cells can, theoretically, still thrive with reduced levels of tau. The nerve cells can still rely on other microtubule-associated proteins which can also carry out most of the cellular tasks that tau performs. Still, as with treatments targeting amyloid secretases, reducing tau levels again raises the issue of 'meddling' with the body's healthy physiology. We already know this is tricky to do, as there are whole cell systems dependent on proteins such as tau, and we might not wish to cause adverse side-effects. As we saw with beta-amyloid, it is better instead to focus on changing the actual process of the disease, such as decreasing the phosphorylation of tau.

Decrease of tau phosphorylation

Let's remember that once tau gets phosphorylated it struggles to carry out its usual role, providing structural support to the microtubules and transporting molecules along the microtubules. The rationale for this treatment is therefore to decrease tau phosphorylation.

There are various ways to reduce the phosphorylation of tau. One particular treatment targets the proteins producing phosphate – phosphatase. By reducing levels of phosphate surrounding tau, this should then lead to less phosphorylated tau. Some of these phosphatase-modifying drugs are already available for other diseases. One medication which has a phosphatase-reduction function is already available for Alzheimer's disease – memantine. Memantine is a drug commonly prescribed in the later stages of Alzheimer's disease and frontotemporal dementia and has been shown to slow down the later development of the disease. However, we need to make it clear that memantine is not primarily a phosphate-reducing medication, but rather that one of its side-effects is removing phosphate and its effectiveness in reducing Alzheimer's disease is still being investigated.

Other treatments targeting tau phosphorylation focus more on how phosphate attaches to tau. Even in healthy cells, phosphate cannot simply attach itself to tau. Instead, it needs the help of another class of proteins called kinases for the phosphorylation of tau to occur. So, targeting how kinase functions can potentially change the rate of tau phosphorylation. As with the phosphatase-targeting drugs, kinase-targeting medication already exists. However, none of these medications has been used in Alzheimer's disease. Instead, most of these drugs treat cancer. It might seem a bit strange to use a cancer medication for Alzheimer's disease, as these are diseases with completely different underlying disease processes. However, why reinvent a medication for kinase function for Alzheimer's disease if we already have an existing medication for cancer, which is – importantly – already approved for human use and readily available?

This kind of medication 'repurposing' is based on the notion that if we have medication treating a specific process in a cell, it may work for many diseases, not just the one it was developed for. Indeed, there are several ongoing studies investigating whether 'repurposing' medication can potentially speed up and help people with Alzheimer's disease.

Decrease phosphorylated tau accumulation

Decreasing the accumulation of phosphorylated tau is a very similar treatment to decreasing beta-amyloid accumulation. For this, we aren't worried about how much phosphorylated tau is generated, instead we want to stop the phosphorylated tau 'clumping' together, something which makes it even more toxic to the nerve cell and harder to remove.

One leading candidate in the fight to stop tau 'clumping' together is methylene blue. Methylene blue is actually a dye but it's also used as a medication for various conditions. We now know that methylene blue can reduce the accumulation of tau, although there is still some controversy as to whether it also changes the toxicity of the accumulated phosphorylated tau. In a nutshell, methylene blue can reduce the clumping together of tau and it may also make the phosphorylated tau less toxic to the nerve cell. Suffice to say, treatments such as methylene blue, as well as curcumin (which can restrict the build-up of amyloid-beta), have been shown to reduce tau accumulation but their exact effects on the development of Alzheimer's disease are still not fully understood and require more investigation.

Increase phosphorylated tau removal

Finally, we come to the method that increases the removal of phosphorylated tau in a similar way to the one we saw applied to beta-amyloid. Again, they are based on immunisation therapies.

As with beta-amyloid monoclonal antibody therapies, phosphorylated tau monoclonal antibody therapies work by attaching themselves to phosphorylated tau and flag it to the body's immune system for removal. It is too soon to know if these tau monoclonal antibody therapies work, as clinical trials are still ongoing. But monoclonal antibody therapies have shown promising results for beta-amyloid so they may have a significant impact on tau medication development in the near future, too.

PART 3.
SUMMARY

In this part of the book, we have covered the following aspects:

- Proteins are responsible for the development of Alzheimer's disease.
- The two main proteins are amyloid and tau.
- Amyloid is part of the healthy nerve cell and provides vital functions.
- Once amyloid gets broken down, the problems start, with parts of amyloid becoming beta-amyloid.
- Beta-amyloid gets ejected out of the nerve cell as a waste product, waiting to be removed.
- If too much beta-amyloid accumulates outside of the nerve cells, it starts glueing or clumping together.
- The clumped-together beta-amyloid is called beta-amyloid oligomers or plaques.
- It is much harder to remove these oligomers or plaques from the body and therefore they can stay outside of the cell for years or decades.

- The slow accumulation of amyloid was seen for decades as the sole cause of Alzheimer's disease (Amyloid Cascade Hypothesis).
- More recent evidence has shown that it is not only beta-amyloid that causes Alzheimer's disease but also tau.
- Tau is another protein in the nerve cells, important for the structure of the cell but also for transporting 'things' through the nerve cell.
- For an unknown reason, too much phosphate in the nerve cell starts to hinder or stop the function of tau, creating phosphorylated tau.
- Phosphorylated tau can no longer function properly, thereby causing a loss of transport in the nerve cell and a reduction in its stability.
- Phosphorylated tau is also toxic to the nerve cell.
- If the phosphorylation of tau is widespread, the nerve cells start dying and the phosphorylation 'infects' or spreads to neighbouring nerve cells.
- It is both the accumulation of beta-amyloid and the spreading of phosphorylated tau which causes Alzheimer's disease.
- The types of medications used to tackle Alzheimer's disease are varied but work via three main principles:
    - Reduce the amount of beta-amyloid or phosphorylated tau.
    - Avoid accumulation of beta-amyloid or phosphorylated tau.
    - Boost the removal of beta-amyloid or phosphorylated tau.
- The most promising medication trials boost the removal of beta-amyloid using monoclonal antibodies.
- There are ongoing trials using monoclonal antibodies for phosphorylated tau.

PART 4.
GENETICS AND LIFESTYLE

The failure of many clinical trials for new medication, along with the availability of biomarkers to detect the disease earlier, has shifted the scientific community's focus onto pinpointing the risk factors for developing Alzheimer's disease. The rationale behind investigating these risk factors is that if we know what they are, we can change our own behaviour and slow down or even stop the development of Alzheimer's disease.

When some of us hear the phrase 'risk factors', we instantly think of genetic risk. In other words, certain genes we've inherited from our parents may predispose us to develop Alzheimer's disease. However, what we often overlook is that Alzheimer's disease lifestyle choices actually often have a much bigger impact than our genes.

Before we go into the details of genetic and lifestyle risk factors, however, we need to understand a few key principles as to why our body becomes more susceptible to disease, such as Alzheimer's, when we age. Grasping these principles will make it easier for us to understand how risk factors can influence our vulnerability to Alzheimer's disease. One reason why we are more susceptible to diseases is that we simply live much longer than before. When homo sapiens appeared 200,000 years ago, our average life expectancy hovered at around 35-40 years old. Anyone above 40 was considered old. The industrial revolution, improvements in sanitation and medical breakthroughs dramatically extended lifespan. Average life expectancy rose to 45 years in 1900, 60 years in 1930, 70 years in 1955 and, finally, more than 80 years in many countries today. In other words, we have doubled our life expectancy since 1850 over only 170 years, compared to the prior 199,830 years when it hovered at around 35-40 years.

One must thank the technological, hygienic and medical advances for this enormous increase in life expectancy. However, living longer also has its downsides, as our body copes less well with diseases and their risk factors.

How does ageing make us more vulnerable to Alzheimer's disease and which risk factors can explain this?
A key thing to understand is that the longer we live, the more time there is for disease processes to accumulate. Such cumulative changes are changes to our body which happen slowly over a long time, without us being aware – at least at first – that they have happened. As previously mentioned, clinicians often refer to these 'creeping in' changes as 'incipient/insidious' changes. A good example for a cumulative, insidious change is atherosclerosis, when our arteries get clogged up over time because of a high cholesterol consumption. When we have atherosclerosis we can't pinpoint exactly when the clogging up of our arteries started. It might have happened last week, last year or a decade ago without us even knowing.

Simply put, the older we get the more time we have had to build up increasing disease processes. Once we understand this fact, it should become clear why the majority of diseases only happen from middle-age onwards. On top of that, as we age our body is no longer as efficient in getting rid of these increasing disease changes.

Coming back to the example of atherosclerosis, our arteries might have clogged up over all these years or decades, but our body would have coped with this process better when we were younger. The older we get, the less our body can 'clean' those arteries and the more they get clogged, until for some people they become so badly clogged that the arteries get blocked. If such a blockage happens in the brain, we will have a stroke. This means that increasing disease changes (atherosclerosis) can lead to a distinct disease change (stroke). For any disease prevention approach this is an important point to grasp, so it's worth restating: increasing disease changes can lead to diseases.

Why is this important?
It explains why certain risk factors in our life (for example eating too much fatty food and doing little exercise) can lead to increasing changes (atherosclerosis), which finally in older age can lead to distinct and potentially fatal diseases (stroke). There is a clear chain of events: eating too much fatty food, doing little exercise atherosclerosis stroke. This might seem frightening but fret not, the good thing is that with lifestyle changes (and medication) we can actually avoid or alter this chain of events, even when we are older.

That's all very interesting in terms of cardiovascular disease but what has it to do with Alzheimer's disease?
It is a very similar case for Alzheimer's disease, but a bit more complicated. Cardiovascular disease provided a good analogy to understand the principles of increasing and discrete disease changes, but the same principles apply to Alzheimer's disease. We already know that the accumulation of amyloid and tau is an increasing disease process, which leads to the discrete disease of Alzheimer's. We also know the chain of events which leads to the accumulation of amyloid and tau accumulation from Part III of the book. However, we also need to understand that the amyloid and tau disease changes can be influenced by other risk factors which can hasten or slow this accumulation.

In a way, we can regard risk factors as the environment/factors in which the disease processes happen. Certain environments or factors will accelerate the process for Alzheimer's disease, while other environments or factors might slow down these processes. This comparison should make it clear that influencing the background mix or milieu of risk factors can make a significant difference to the development of the disease, since we can either speed up or slow down the accumulation of amyloid and tau. This brings us to the last area of risk factors: their division into modifiable and non-modifiable risk factors.

Modifiable means we can change those risk factors, mostly with our lifestyle choices or with existing medication. Modifying those risk factors changes the environment or factors for the accumulation of amyloid and tau, which in turn increases or decreases our risk for Alzheimer's disease. In other words, if we make the 'right' lifestyle choices we can reduce our risk of developing Alzheimer's disease by up to 40%.

Non-modifiable risk factors are far trickier, since by definition we cannot change the risk associated with them. Still, it is important to know about such non-modifiable risk factors for Alzheimer's disease, but we should not worry too much about them since we cannot change them. Instead, it is more important to focus on the modifiable risk factors that can potentially delay or even avoid Alzheimer's disease.

Based on this, I have divided the chapter on genetic and lifestyle risk factors into modifiable and non-modifiable risk factors. As the old saying goes: 'It's not about the cards you're dealt, but how you play the hand.'

30.
GENETICS 101

In my experience, many people are concerned that they might have inherited genes predisposing them to developing Alzheimer's disease, as their mother or father or close relatives have or had some form of dementia. Indeed, I can myself name a few relatives of mine who had dementia. The good news is that it is unlikely that we have inherited genes from our parents or relatives that make it certain that we will develop Alzheimer's disease. The scientific evidence is that less than 10% (10 in 100) cases of Alzheimer's disease are related to genetics and only 1% (1 in 100) are caused by our genes. We will explain the difference between related to and caused by later, but suffice to say the chances of us having inherited genes that lead us to develop Alzheimer's disease are low.

If this is the answer you were seeking, you can relax and enjoy the remainder of Part IV during which we will explore the science of genetics in these rare cases of genetic Alzheimer's disease. In addition, we will explore genes which can increase or decrease our overall risk for Alzheimer's disease.

Ready for the deep dive into the weird and wonderful world of genetics?
Here we go.

Genetics is a vast galaxy unto itself within the universe of biology. Given its sheer size, we can only 'dip our toes' into its waters when trying to improve our understanding of the genetics of Alzheimer's disease. We will, therefore, focus solely on the aspects of genetics that are relevant to the development or risk of Alzheimer's disease.

Let's start with the basics: what are genes?
Genes are simply the instructions or blueprints for the production of proteins in our body. Proteins are vital for the healthy functioning of our body and when they go wrong, proteins such as amyloid and tau can cause havoc and potentially devastating diseases. Their blueprints are clearly of great importance for the body. Because of this, we have two sets of each gene. You might have learnt at school that our DNA, which carries our genes, comes in the form of a double-helix structure with two strands wound around each other. The two strands of genes are called a chromosome. We inherit each chromosome from our biological parents, one from our mother and one from our father. Although the two chromosomes are very similar, they are not completely identical, and the minute differences between them determine many factors, such as what eye or hair colour we have. This is because only one chromosome is 'read' to produce the blueprint for the proteins.

If we have two non-identical copies of each gene, how does the body know which one to read?
To answer this question, we need to understand another fundamental aspect of how genes work. Some genes are dominant while others are recessive (scientific speak for submissive). Importantly, dominant genes will always try to produce (or in genetic speak, 'express') their proteins. The dominant gene may have been inherited either from our mother or father but it always produces its proteins, regardless of the other inherited gene on the other strand. For example, the way in which we inherit freckles from our parents is a good example of a dominant gene inheritance. That we have freckles is largely determined via one dominant gene (MC1R – geneticists love using unpronounceable acronyms, so let's just enjoy them), which we inherited from either our mother or father. If either of our parents

has this gene and we inherit from them, then it is highly likely that we will also have freckles.

What about if we inherit this gene from both of our parents?
Then your chances of having freckles are even higher, as you will have two dominant genes, one on each genetic strand, both of which will try to produce as much protein as possible. However, notice that we use the word 'likely', which is important because genetics is largely probabilistic. This means that even if we have a dominant gene(s) it is not 100% certain that this gene will be produced. The reasons for this likelihood are quite complex, so for now let's simply acknowledge that it is so. The reason why we need to bring up the likelihood of gene expression is that geneticists use a different term for the probability that a gene will be active and produce a lot of proteins – penetrance.

A high penetrance means that it is very likely that a gene will be highly active, while a low penetrance means that the gene will be less active. From what we now know, it should become clear that dominant genes have a high penetrance. That's because even if we only have one copy of this gene on one of our genetic strands, it will likely produce the proteins it has the blueprints for. By contrast, low-penetrance genes are mostly recessive genes. We know now that the likelihood of a gene being active and producing proteins is much lower for low-penetrance genes, such as recessive genes. This explains why for recessive genes to be active and produce proteins, we usually need to inherit the same gene from both of our parents. Inheriting a recessive gene from one parent is – mostly – not enough for the gene to be active and to produce proteins. Or if the recessive gene is active, it shows only very low activity, with little protein produced.

A good example of recessive genes is how we inherit our eye colour. Our eye colour is determined by the pigmentation pattern of the iris in our eyes. One of the main genes involved in the pigmentation of our iris is OCA2, which, if highly active, will result in a blue eye colour. Since OCA2 is a recessive gene, we need to inherit two copies of it, one from each of our parents, to have blue eyes. If we inherit only one copy from one parent, then we will have brown eyes, as the genes for brown colour pigmentation for the iris are more dominant. Remember that we only need one copy of a dominant gene for it to be dominant, while we must have two recessive genes for them to be active. That's why blue eyes are so much less common in the population than brown eyes, as we need to inherit the recessive genes for blue eye colour from both of our biological parents. We now have some fundamental knowledge of how genes work and what factors determine how active they are and, in turn, how many proteins they produce. Let's apply our new knowledge to Alzheimer's disease and how genes can influence our risk of the disease.

31.
FAMILIAL ALZHEIMER'S DISEASE

Let's start off with the most difficult topic – genetic changes which are not modifiable by our lifestyle or medication. These non-modifiable genetic types of Alzheimer's disease are called 'familial Alzheimer's disease'. All these familial cases occur because there is a mutation in one of the genes that we inherit from our parents that causes us to develop Alzheimer's disease.

A mutation is, literally, a change in the genetic code (from Latin mutare, meaning 'to change'). The mutation causes a change to the blueprint for the proteins which causes the proteins to 'behave' differently. Such mutations may occur in our genetic material in many different ways. The exact reasons why these mutations only happen in certain families is still being investigated. The good news is that such gene mutations for Alzheimer's disease are extremely rare. To date, this kind of mutation accounts for less than 1% of all Alzheimer's disease worldwide which means that it is very unlikely that we have 'inherited' Alzheimer's disease from our parents. The genetic mutations that cause Alzheimer's mostly affect the following three genes: APP, PSEN1 and PSEN2. Since we already know that the accumulation of beta-amyloid is particularly relevant to Alzheimer's disease, it should come as no surprise that mutations to all three genes (APP, PSEN1 and PSEN 2) impact the production or accumulation of beta-amyloid, resulting in higher levels of beta-amyloid. Let's go through each one now in detail to understand how the mutation influences the production of beta-amyloid.

APP

APP is the abbreviation for the Amyloid Precursor Protein gene, which can be found on chromosome 21 of our DNA. The role that APP plays in the development of Alzheimer's disease has been known since the 1980s, when it was discovered that people with trisomie 21 – a triple copy of chromosome 21 (more commonly referred to as Down syndrome), develop Alzheimer's disease-type changes in their brain. I deliberately say 'Alzheimer's disease-type changes', as we need to make clear that Down syndrome is not Alzheimer's disease.

However, since people with Down syndrome have three copies of chromosome 21, they have triple the production of amyloid in their brain. This means that people with Down syndrome show beta-amyloid oligomers and plaques in their brain, often in their 40s or 50s, and many start to develop Alzheimer's disease later on in life. The discovery that three copies of APP on chromosome 21 leads to higher beta-amyloid production was therefore the first indication that genetics might play a role in the development of Alzheimer's disease in some people. Scientists then started to 'look for' families which might have changes in the APP gene on chromosome 21 and, if so, whether this changed their risk for Alzheimer's disease. The first family with an APP mutation was discovered in 1991, confirming the hypothesis that APP is important in the development of Alzheimer's disease.

In this family it was found that they had something called a missense mutation in the APP gene. A missense mutation is a mutation where, for some reason, only a single letter of the genetic code changes. But this single letter change can have devastating consequences, as the proteins produced by this gene are therefore faulty and cannot perform their normal function. In the APP gene mutation, this leads to a significant increase in beta-amyloid.

For the family in question, the single letter genetic code change in the APP gene apparently occurred in the grandfather of the family, and he passed on this mutation to his two sons and many of their children. It is not clear whether the mutation occurred for the first time in the grandfather because dementia was not generally recognised and diagnosed prior to that time. So, it's equally possible that the family may have had this mutation for generations or that it was new in the grandfather. Either way, it is typical for the genetic mutations that cause Alzheimer's disease to affect whole families – hence again the name 'familial Alzheimer's disease'. Since the dominant gene is inherited from one parent, we can often track the disease back through the family as only people with the affected gene in the family will have the disease.

Therefore, a common question arises: how many family members need to have been affected for you to be at risk of having such a gene mutation? Most of us have someone in our families who has, or had, dementia or Alzheimer's disease. Does this mean that you could have inherited a gene from them which increases your risk for Alzheimer's disease?

How many of our relatives need to have been affected by dementia or Alzheimer's disease in order for us to be considered for genetic testing is unclear. However, for a doctor to recommend genetic testing a patient usually needs to have many first- and second-degree relatives affected by dementia. For the vast majority of us, this will not be the case as we are likely to know very few – if any – members of our family who have been affected by Alzheimer's disease – myself included. For example, maybe your mother or father had dementia and maybe an uncle or aunt. This is completely normal and does not suggest a form of familial Alzheimer's. By contrast, for families with familial Alzheimer's disease, many if not most of all family members will have been affected by the disease.

Let's now explore how the other common type of familial Alzheimer's disease, affecting the PSEN genes, causes the disease.

PSEN

PSEN stands for presenilin. The genes for presenilin (PSEN1 and PSEN2) can be found on chromosome 14 and 1 of our DNA, respectively. Mutations in both proteins are extremely rare, with fewer than 0.5% of all Alzheimer's disease cases worldwide affected by such mutations. PSEN genes are important for the production of the gamma secretase. Let's remember that the gamma secretase acts like a scissor snipping amyloid into beta-amyloid. The exact mutations caused to the PSEN genes are well described. Most of them are again missense mutations, where one letter of the genetic code for PSEN has been changed. However, what is still not entirely clear is how these mutations change the 'behaviour' of the gamma secretase.

Currently, there are two main possibilities. The first possibility is that the PSEN mutations increase the activity of the gamma secretase, so that overall more beta-amyloid is produced. Since more beta-amyloid means an increased risk for Alzheimer's disease, this makes sense. However, a second possibility suggested by some studies is that the PSEN mutations change the gamma secretase so that it produces a type of beta-amyloid which is more prone to accumulate. This would mean that PSEN mutations do not increase beta-amyloid production, but instead produce a type of 'stickier' beta-amyloid. This stickier beta-amyloid is more likely to form amyloid oligomers and plaques, which we know increases our risk of Alzheimer's disease.

However they work, PSEN mutations on the gamma secretase have a devastating effect on families, as it is nearly 100% certain that all family members carrying either a PSEN1 or PSEN2 mutation will eventually develop Alzheimer's disease. We have already mentioned that if families have dominant gene mutations, such as APP or PSEN1 or PSEN2, multiple members of the family are usually affected, since we only need one copy of the mutated gene for Alzheimer's disease to develop. There is, however, another sign that a gene mutation for Alzheimer's disease is within our family's genetic pool – the age at the onset of the disease.

These mutations speed-up the production or accumulation of beta-amyloid which often means that people with these mutations develop Alzheimer's disease at a much younger age. Even younger people with PSEN mutations have unusually high levels of beta-amyloid in their brains. Unsurprisingly, many people with PSEN mutations develop Alzheimer's disease before the age of 60.

The youngest person with a PSEN mutation whom I have personally met started to show symptoms of Alzheimer's disease when they were in their 30s. But it is not unheard-of for people in their 20s to develop Alzheimer's disease if they have a PSEN mutation. Sadly, older and younger family members alike might be affected at the same time. Genetic investigations for Alzheimer's disease are therefore only often conducted when multiple generations and relatives of a family are affected and the disease starts at a younger age for those affected. It is worth reiterating that these mutations are exceedingly rare. Even if you have had a few people in your family with some form of dementia, it is very unlikely that it was caused by a genetic mutation.

The rarity of such mutations makes it even more astonishing that the most famous case of Alzheimer's disease had a PSEN1 mutation – Mrs Auguste Deter. That she had a PSEN1 mutation was not known at the time Alois Alzheimer saw her. However, in 2013 scientists re-analysed the brain tissue that Alzheimer had used for his analysis over 100 years before. They found that Mrs Deter had a PSEN 1 mutation affecting one letter of her genetic code on chromosome 14. It is astonishing that Alzheimer literally found the needle in the haystack when identifying Mrs Deter, since the prevalence of familial Alzheimer's disease is so low. But retrospectively it makes sense, because Alzheimer noticed Mrs Deter partly because she was so young (51 years of age) when he saw her at the hospital in Frankfurt. At the time, Alzheimer and his contemporaries saw many people who showed signs of dementia at older age, which they called 'senile dementia'. However, Mrs Deter struck Alzheimer as being too young for senile dementia. He therefore often referred to pre-senile dementia in his notes about Mrs Deter; today we would describe a dementia that develops before the age of 65 as 'younger onset dementia'. Still, there is some irony in the fact that the 'discovery' of the disease was actually based on its rarest form, due to a genetic PSEN1 mutation. (Interestingly, the cases which Oskar Fischer described in his publications were all senile dementia and therefore he, in fact, described the more common form of the disease, rather than the exception, as Alzheimer did.

More recent research into the disease has also relied on the generosity of those families affected by familial Alzheimer's disease. The reason being that before biomarkers for Alzheimer's disease were available, cases of familial Alzheimer's disease offered the sole source of data for scientists when investigating the possible origins of the disease and which symptoms would emerge first. For example, among families with APP or PSEN mutations, it was possible to study those members who had a known genetic mutation for Alzheimer's disease but had not yet developed any symptoms of the disease. Such 'presymptomatic' people are incredibly valuable for Alzheimer's disease research because they give scientists the opportunity to study how the disease develops and how symptoms eventually appear.

Despite the importance of familial Alzheimer's disease, there is one downside to relying on research in this area. Familial Alzheimer's disease is caused solely by mutations affecting beta-amyloid, but we now know that both amyloid and tau are important for the development of Alzheimer's disease. This helps to explain why there was such an emphasis on beta-amyloid as the cause of Alzheimer's disease before the availability of biomarkers. In fact, the original Alzheimer Cascade Hypothesis was based on PSEN familial Alzheimer's disease. Focusing purely on amyloid therefore seemed the obvious path for Alzheimer's disease prevention and treatment, but we now know that this approach underestimated the role of tau in the development of Alzheimer's disease.

Speaking of which, do genetic mutations for tau exist and do they cause Alzheimer's disease?

MAPT

The tau protein also has a genetic code providing its blueprint – the microtubule-associated protein tau gene (MAPT) on chromosome 17. However, the link between genetic mutations in MAPT and the development of Alzheimer's disease is much weaker than the amyloid related genes (APP, PSEN1, PSEN2). The reason for this is not clear.

MAPT mutations have been linked to other dementias, such as frontotemporal dementia and other neurodegenerative diseases. However, there is little evidence that mutations in MAPT cause familial Alzheimer's disease. Instead, mutations of MAPT have been shown 'only' to increase our risk of the disease. The distinction between having familial Alzheimer's disease or only an increased genetic risk for Alzheimer's disease leads us back to the start of the chapter when we talked about penetrance of genes. We already know that amyloid-related mutations (APP, PSEN1, PSEN2) have a high penetrance, meaning that anyone with those genetic mutations will most definitely develop Alzheimer's disease. However, for MAPT mutation there seems to be a low penetrance for Alzheimer's disease. This means that even if we have a genetic mutation in MAPT it will 'only' increase our risk for Alzheimer's disease but it is far from certain that we will actually develop Alzheimer's disease as a result of this mutation.

It is important to understand this distinction before we explore other 'risk genes' for Alzheimer's disease in the next section. Similar to MAPT for Alzheimer's disease, such risk genes can increase or decrease our risk. However, having such risk genes does not mean that we definitely will develop Alzheimer's disease.

32.
'MODIFIABLE' RISK GENES

A variety of so-called 'risk genes' can increase or decrease our risk of getting Alzheimer's disease. Now these risk genes do not not actually directly influence amyloid or tau levels. Instead, they often change the environment (or what scientists call the milieu) for amyloid and tau. If the risk gene encourages a milieu which is conducive to the accumulation of amyloid or tau, it increases our risk of Alzheimer's disease. By contrast, if the risk gene allows a milieu which is not conducive to accumulation of amyloid or tau, it decreases our risk.

This means that risk genes can be either good or bad for the development of Alzheimer's disease. Even more important is the realisation that, since these risk genes only influence the milieu for amyloid and tau accumulation, we can try to influence the milieu ourselves, either by taking medication or by making lifestyle choices. So even if we are at an increased genetic risk, we can dampen or ameliorate that risk through medication or lifestyle choices. In other words, we can determine how bad for us those genes actually are. The extent to which we can influence our genetic risk is critical for distinguishing familial Alzheimer's disease genes and risk genes. For familial Alzheimer's disease the milieu matters very little as the genes themselves 'drive' the accumulation of amyloid because of their high penetrance. However, for risk genes the milieu is important and it is within our influence. It means that we have control over our risk genes and can regulate them through medication and lifestyle choices. That is not the case for familial Alzheimer's disease mutations.

What are these risk genes then and how many are there?
There are now ~ 30 known genes which increase or decrease our risk for Alzheimer's disease. The vast majority of these are very rare and only minimally increase the risk for Alzheimer's disease. So there is little point in discussing them. Instead, we will focus on two risk genes which are more common and can influence our future risk of Alzheimer's disease – APOE and TREM2.

APOE
APOE stands for the 'Apolipoprotein' gene, which can be found on chromosome 19 of our DNA.

Why is APOE relevant as a genetic risk factor for Alzheimer's disease?
We don't exactly know: the role of why certain variations of APOE increase our risk for Alzheimer's disease is still being investigated. What we do know is that certain variations of APOE can increase our risk of Alzheimer's disease by a factor of 12.

What do we mean here by 'certain variations' of APOE?
To explain what 'certain variations' mean, we have to introduce another genetic term – genetic polymorphism. Genetic polymorphism (from Greek 'poly' = many; 'morph' = form, meaning literally 'many forms') refers to having different variations of the same gene in the population.

A polymorphism is a change in the DNA commonly found in the population. This means that different people in the population have different forms of the gene. For APOE, we have several genetic polymorphisms, with the most relevant for Alzheimer's disease being e3:e3; e3:e4 and e4:e4. The e stands for the particular allele carrying the polymorphism; the number stands for the type of polymorphism. Since we have two alleles (remember DNA has two strands) we can have different combinations or variations of APOE. Does the APOE genetic polymorphism really matter for our risk of Alzheimer's disease?

It matters a lot, as the type of APOE polymorphism determines the level of future risk for developing Alzheimer's disease. The most common APOE polymorphism is e3:e3. Around 78% of the population has the e3:e3 APOE polymorphism. It is regarded as the 'normal' or standard APOE polymorphism. Importantly, the e3:e3 polymorphism neither increases nor decreases our risk of developing Alzheimer's disease later. Around 12% of the population – one in eight – have the e3:e4 polymorphisms, which increase our risk of Alzheimer's disease threefold. This means that we are three times more likely to develop Alzheimer's disease, compared to someone who has the e3:e3 APOE polymorphism. Let's also remember that having a higher genetic risk often means that we will develop the disease earlier, as the accumulation of amyloid and tau is accelerated. Twelve per cent have the e3:e4 polymorphism and develop Alzheimer's disease on average up to 10 years earlier than someone who had the 'normal' e3:e3 polymorphism.

One in fifty people – 2% – have the e4:e4 polymorphism which increases our risk for Alzheimer's disease 12-fold, meaning that we are 12 times more likely to develop Alzheimer's disease compared to someone who has the e3:e3 APOE polymorphism. Having the e4:e4 polymorphism is therefore far more concerning as we are at a much higher risk of developing Alzheimer's disease than the rest of the population. People with e4:e4 develop Alzheimer's disease up to 20 years before people who have e3:e3. Sharp-eyed readers might have spotted already that the percentages do not add up to 100% (e3:e3 = 78%; e3:e4 = 12%; e4:e4 = 2%).

Who are then the remaining 8% of APOE polymorphisms?

They are the e2:e2 and e2:e3 polymorphisms, which reduce our risk of Alzheimer's disease by up to 40%, compared to the e3:e3 'average'. Inheriting our genes is therefore a bit of a lottery, as by sheer luck some of us will have inherited better or worse genes for Alzheimer's disease.

Where we live and our ethnic group also influences our risk of getting Alzheimer's disease. Indeed, the risk of APOE polymorphisms on the development of Alzheimer's disease varies across different ethnic and country populations. For example, e4:e4 increases the risk of Alzheimer's disease in African Americans 'only' six-fold and in North American Hispanics only two-fold, compared to 12-fold in Caucasian Americans and an astonishing 33-fold in Japanese people, meaning a 33 times higher risk. We have to accept that there are large differences in how much the genetic polymorphisms increase our risk of Alzheimer's disease, according to which ethnic group we belong to. The exact reasons for this are not yet clear. It might be due to environmental factors influencing our genetic risk; alternatively, it could also be that APOE's functioning varies slightly in different ethnic groups.

How then does APOE increase our risk for Alzheimer's disease?

Exactly how APOE impacts our risk of developing Alzheimer's disease is yet to be determined. However, we know that the different polymorphisms of APOE seem to be involved in the accumulation and clearance of beta-amyloid in the brain. APOE itself does not interact with the beta-amyloid but it creates a milieu which is either more conducive to the removal of beta-amyloid (e2:e3 and e2:e2 polymorphisms); or hinders the removal of beta-amyloid (e3:e4 and e4:e4 polymorphisms); or has no effect on the removal of beta-amyloid (e3:e3 'average' polymorphism).

In summary, APOE's various roles create a milieu which can affect our future risk of developing Alzheimer's disease. Finally, let's remind ourselves again that the 'increased genetic risk' due to APOE does not mean that we will develop the disease, but it can increase or decrease our risk for Alzheimer's disease. The same applies for the other common Alzheimer's disease gene – TREM2.

TREM2

TREM2 is a gene located on chromosome 6 of our DNA. TREM2 stands for – brace yourself – 'Triggering Receptor Expressed on Myeloid cells 2' gene. This rather strange name is explained by the fact that scientists are fond of describing things in a very literal way.

For TREM 2 this means that the gene produces a protein within myeloid cells. Myeloid cells are found mainly in our bone marrow and produce proteins which trigger an inflammatory response to injury or disease.

What does a bone marrow cell protein have to do with Alzheimer's disease?

TREM2 helps regulate the inflammatory response, in our brain as well as our body. We have already explored how our immune system can help remove beta-amyloid or phosphorylated tau. Inflammation is part of this immune system response. Basically, the increased accumulation of beta-amyloid creates an inflammatory response to activate our immune system to deal with the increase. Inflammation can be seen therefore as a signal that the immune system is actively trying to get rid of beta-amyloid.

In the brain, TREM2 is often found in the microglia cells which play a key role in the inflammatory or immune response in the brain. For example, if there is too much beta-amyloid outside of the nerve cells, it triggers the microglia cells to spring into action by starting a process called phagocytosis (from Greek, meaning 'phago' – to eat and cytosis – of a cell). Phagocytosis destroys potentially dangerous material by letting the immune system 'eat' the threat, like beta-amyloid. Over the recent years it has been found that TREM2 levels are elevated in corticospinal fluid in people with Alzheimer's disease. The elevated TREM2 levels indicate that there is an inflammatory process happening in the brain, because of the accumulating beta-amyloid.

However, sometimes a gene change dampens down this inflammatory response. This is what can happen if TREM2 changes/mutates.

TREM2 mutations increase our risk of Alzheimer's disease two- to four-fold, meaning we are two or four times more likely to develop Alzheimer's disease later in life compared to the normal population. As you might remember from the APOE chapter, this is a similar increase in risk as having e3:e4 alleles for APOE. TREM2 is a rarer Alzheimer risk gene which, in essence, diminishes our body's immune system response to the accumulation of beta-amyloid.

How TREM2 mutations lessens the immune response is still being explored. But it has become increasingly clear that Alzheimer's risk can be strongly modulated by our inflammatory response, making a big difference to the development of the disease.

33.
A WORD ON GENETIC TESTING

Since we now know that genes can either determine that we will definitely develop Alzheimer's disease (APP, PSEN1, PSEN2) or increase our risk for Alzheimer's disease (for example APOE, TREM2), why don't we test everyone for their genetic risk?

Such a genetic risk screening would enable us to not only identify people early enough but also give them an opportunity to modulate their future risk of Alzheimer's disease. Technically, it is now very easy to get our genes tested for such 'risk' genes. Several private gene testing companies offer services for having our risk genes (APOE) for Alzheimer's disease tested. But before we rush off and test ourselves, let's ask ourselves, do we really want to know our genetic risk for Alzheimer's disease?

For some people the answer is clear – yes, they want to know their genetic risk. If they are at an elevated genetic risk of getting Alzheimer's disease, they can then change their lifestyle to potentially reduce that risk, or if they are not at an elevated risk they can worry less about Alzheimer's. For others, however, genetic testing opens 'Pandora's box'. For example, we might have to disclose genetic results to our healthcare system or insurer or even to our employer. This might not be a problem for countries with a national health system, like the UK, however for countries or people with private health insurance this information can have a significant impact on their insurance premium or even employment. More importantly, we may also need to disclose this genetic information to our family and particularly to our children, since we share our genes with our family. This can be particularly challenging, as our genetic information not only affects us but anyone biologically related to us.

Technically, it is therefore very easy to get ourselves tested for a genetic risk for Alzheimer's disease but ethically there are significant considerations to be taken into account before taking this step. My suggestion is that if you are really worried that you have an increased genetic risk for Alzheimer's disease, you should first discuss this with your family doctor. If you and your doctor decide that it might be worth investigating then you would usually be referred to a genetic counselling service. Genetic counselling services specialise in dealing with genetic diseases and can advise you whether genetic testing would be appropriate in your case and what potential implications it might have for your life and family.

34.
NON-MODIFIABLE LIFESTYLE FACTORS

There is an irony that many of us are concerned about our genetic risk for Alzheimer's disease, but less worried about how our lifestyle choices might impact our risk for the disease. The irony lies in the fact that we now know that our genetic risk for Alzheimer's disease is in fact small. The scientific data is clear that our lifestyle has a much bigger impact on our future risk for Alzheimer's disease.

Scientific evidence has shown that up to 40% of risk for Alzheimer's disease is down to the way we live. This is an astonishing figure which usually makes people sit up, as we are not generally aware how much of a difference our lifestyle choices can make for the future development of disease. Similar to genetic risk factors, our lifestyle factors can be modifiable or non-modifiable. Let's start first with the non-modifiable life factors, things which we cannot change.

Why talk about non-modifiable risk factors if we cannot change them?
The main reason is to know and understand these risk factors so at least we know what they are. Let's start with the top non-modifiable risk factor for Alzheimer's disease – age.

Age

It is almost too obvious to state that age is the main non-modifiable risk factor for Alzheimer's disease but it is important to understand this fact. The obvious reason why age is the top non-modifiable risk factor for Alzheimer's disease is that our risk for the disease increases the older we get. Scientific evidence is clear that our risk for Alzheimer's increases with age. Only about 2% (2 out of 100) people develop Alzheimer's between the ages of 65 to 69. However, 19% (19 out of 100) people develop Alzheimer's between the ages of 85 to 89.

Since we cannot change the simple fact that we age – at least not yet – this remains the top non-modifiable life risk factor for Alzheimer's disease. Simply put, the longer we live the higher our risk of developing Alzheimer's disease. The reason, as we know by now, the older we get the more time there is for amyloid and tau to accumulate and hence we have a higher risk of Alzheimer's disease in older age.

Education

The second most common non-modifiable risk factor for Alzheimer's disease is education, specifically formal or school education in our early life. The scientific evidence is that the more years of education we had in early life, the lower our risk for developing Alzheimer's disease later on in life. Higher levels of education decrease, therefore, our risk for Alzheimer's disease.

How can we explain this surprising finding?

One potential reason for education being considered a risk factor for the development of Alzheimer's disease is that the symptoms of the disease are measured by cognitive tests, for example to see how well our memory is functioning. If we are performing badly on cognitive tests, it is an indication that we might have the first signs of Alzheimer's disease. However, if we perform well on the cognitive tests then we seemingly do not have Alzheimer's disease. The critical point is that having more formal education means that we are generally more experienced doing tests and might perform better. That is exactly what several studies have found that people who had more years of formal education behind them, but who also had biomarker-proven Alzheimer's disease changes in their brain, still did better on cognitive tests than people with less formal education.

Formal school education gives people more experience of doing tests, of course, but it and also seems to improve a phenomenon called 'cognitive reserve'. One can think of cognitive reserve as our brain's ability to overcome cognitive challenges. We all face cognitive challenges in our everyday life but 'cognitive reserve' refers more to how our brain deals with the cognitive challenges when we have a brain disease. Cognitive reserve allows us to maintain most of our cognitive function, even when a disease takes hold in our brain. But, cognitive reserve does not last forever and depletes over time – it is metaphorically and literally a 'reserve'. Cognitive reserve therefore allows us to maintain our cognitive function at a higher level, even when the process of Alzheimer's disease has started in our brain. However, once the cognitive reserve is depleted, our cognitive function will quickly decrease to a similar level of people who had less cognitive reserve.

Does it therefore matter if we have more cognitive reserve, if we ultimately end up at the same place than people who have less cognitive reserve?
It does, as our cognitive function determines a lot of our daily functioning, such as dealing with finances, cooking meals and generally caring for ourselves. If we have higher cognitive reserve, we are likely to keep those everyday functions for longer, which in turn allows us to live more independently for longer, even if we are in the early stages of Alzheimer's disease.

There is clearly a controversial element in stating that formal school education, or the lack of, is a risk factor for Alzheimer's disease. For one, there is very little we can do to change our formal school education years when we are older. It is therefore a true non-modifiable risk factor but seems quite deterministic, meaning that this factor, which we cannot change, might determine our risk of Alzheimer's disease in later life.

How about missing formal education but engaging in life-long learning and being intellectually very active all your life? Would these not boost your cognitive reserve as well?

These are fair questions to which there are currently no scientific answers but many studies are under way to investigate whether these factors make a difference in cognitive reserve. This would be important to know, since if there is evidence that lifelong intellectual activity builds up our cognitive reserve, it would mean that this risk factor might not be as 'non-modifiable' as we currently think. We can then design interventions to build up our cognitive reserve throughout life, or even when the disease starts.

This has been referred to as cognitive fitness, akin to physical exercise, since we would be able to train your cognitive health. However, be very careful with programs or apps promising cognitive training improvements. So far the scientific evidence that they work is, at best, mixed. The main issue with such so-called cognitive training programs is that we get better at the task they give us but there is little evidence that it actually changes our cognitive performance in real life, not to mention our cognitive reserve. We don't yet know what might be the best cognitive training to prevent or lessen the impact of Alzheimer's disease.

Hearing loss

The final non-modifiable risk factor is also the least understood – hearing loss. Hearing loss was another somewhat surprising risk factor for Alzheimer's disease, even for the scientific community. The scientific data indicates that having hearing loss in middle-age is a risk factor for the subsequent development of Alzheimer's disease.

It's quite normal to lose hearing as we age. It can be caused by several reasons, including long exposure to sounds which are either too loud or too long, medication or other health conditions, such as high blood pressure or diabetes. It is therefore unclear whether hearing loss really is a risk factor for Alzheimer's disease or a collateral effect of another risk factor for Alzheimer's disease, such as high blood pressure (which we will explore in the next chapter). Another element is whether the hearing loss leads to greater social isolation, as people with hearing loss often engage less socially because doing so is harder.

There is clearly more research needed to explore this risk factor, as virtually all studies showing hearing loss as a risk factor for Alzheimer's disease have looked at existing data. However, there is potentially some good news on this, since using a hearing aid has been shown to reduce future risk for dementia. It is, therefore, people with hearing loss who do not use a hearing aid who are at higher risk of dementia later in life. If you have hearing loss and use your hearing aid as prescribed, your risk for future dementia is not increased. These findings suggest that this risk factor might be actually more modifiable than previously thought, since we can modify our dementia risk by using a hearing aid.

Let's move now to the modifiable lifestyle factors to reduce our dementia risk.

35.
MODIFIABLE LIFESTYLE FACTORS

To quickly re-cap, modifiable risk factors that we can actively change can reduce our risk for Alzheimer's disease by up to 40%. This is clearly great news. However, there are so many modifiable risk factors that it can be hard to choose which ones to focus on. As so often in life, breaking it down to a few components can help us achieve our goals. So let's focus on the top four modifiable risk factors, which will make the biggest difference, not just to reduce our risk for Alzheimer's disease but to ensure we age healthily.

Diabetes, hypertension and obesity

We are going to explore these three risk factors together, as they often go hand in hand.

Obesity is commonly estimated by our Body Mass Index (BMI), which is our body weight in kilograms divided by our height in metres squared. So BMI gives a rough estimate of our weight in relation to our height. It is a practical, though imperfect, measurement to provide a quick indicator of a person's ideal weight. Current health guidelines for adults propose that a healthy BMI score lies between 18.5 and 24.9. If our BMI is below 18.5, we are considered underweight. Conversely, if we have a BMI between 25 and 29.9 we are considered overweight. A BMI of 30 to 39.9 is considered obese and a BMI of above 40 morbidly (severely) obese.

Most populations worldwide are moving towards being overweight and obese. This is mostly due to the wide availability of cheap, highly calorific food amid falling levels of physical activity. For example, UK data shows that 28.7% of adults are obese (US 36%; Australia 32%; Germany 19%). This means that almost two-thirds of the UK population, 64.3%, is overweight or obese (US 69%; Australia 66%; Germany 54%) and the trend is that this percentage will increase.

Having a few extra pounds might not mean much when we are younger, but it can make a big difference to our health as we age. If we are overweight or obese, our risk of developing several age-related diseases increases significantly. The reason for this is that the excess fat makes our body less able to cope with other health challenges. A particular worry is that obesity will lead to Type II Diabetes, which stresses our body and increases our risk of developing other health conditions.

Being obese also means that our cardiovascular health worsens as well since our heart needs to pump blood through a much bigger body. This is not just a problem if we are physically active as the activity trains our heart to work harder to supply our newly gained muscles and body mass. However, when we are obese, we are less likely to be physically active, which means our heart has to deal with a larger body without having been trained to supply it. To deal with the larger body, our cardiovascular system commonly takes a shortcut and increases our blood pressure to maintain the supply of blood. In the long-term, this increased blood pressure can result in hypertension. Long-term hypertension can cause a variety of future health conditions, including strokes and dementia.

The danger of obesity is therefore not to be underestimated, as it can lead to severe health conditions later in life and can even cut our life short by a few years.

What has all this to do with Alzheimer's disease?

We already know that hypertension can cause strokes in the brain. Such strokes, and particularly mini-strokes, significantly increase our risk of Alzheimer's disease. Obesity can also lead to pre-diabetes and metabolic syndrome. Metabolic syndrome results in poor regulation of insulin – which regulates our blood sugar – and is often also referred to as insulin resistance syndrome. Metabolic syndrome can cause our blood sugar levels to surge dangerously, putting stress on our body. Insulin plays a role in the accumulation and clearance of amyloid in the brain. Exactly how insulin supports the clearing of amyloid is still being investigated but higher levels of insulin in our body are associated with less efficient clearance of brain amyloid. We already know that high levels of amyloid increase our risk for Alzheimer's disease and therefore the insulin changes caused by obesity and the associated metabolic syndrome can further increase our risk for the disease.

Currently, there is no direct evidence that insulin resistance affects tau phosphorylation. However, people with Type II diabetes have been shown to have higher levels of phosphorylated tau in their corticospinal fluid. Type II Diabetes is the more common form of diabetes and develops for most people in middle-age as a result of excessive weight or sugar consumption. (Type I diabetes is present from birth and, thus far, has not been shown to be associated with a higher risk for Alzheimer's disease.) Since what we eat and how much we move can affect its onset, Type II diabetes is often referred to as a lifestyle disease. The good news is that Type II diabetes can be prevented by eating healthily, being physically active and not being obese. However, even if we have developed Type II diabetes, we can manage our risk, as medication can control our insulin levels. Scientific evidence has shown that people on well-regulated insulin treatment do not show an increased risk of Alzheimer's disease. The risk is highest for people who have undiagnosed Type II diabetes or when people's insulin levels are poorly regulated. As always, prevention is better than treatment, so if we can avoid developing Type II diabetes in the first place, all the better.

Based on the significant impact that obesity can have on our risk of Alzheimer's disease, it is paramount that we keep an eye on our weight as we age. It is fine to carry a few extra pounds when we get older, however, if possible we should avoid weight changes which make us obese or morbidly obese. Once our weight crosses the obese threshold, our risk for Alzheimer's disease – and many other age-related health conditions — increases significantly. As always in life, it all adds up.

Talking about weight leads us nicely into the next risk factor for Alzheimer's disease – lack of physical activity.

Physical activity

Physical activity is not only vital for our muscles and heart, but also for our brain health. There is an abundance of scientific evidence showing that increased physical activity improves our brain health while at the same time reducing our risks for many diseases. Governments all over the world are encouraging people to be more physically active in their everyday lives. But the opposite is the case, with surveys showing that we are now 20% less active than we were in the 1960s – at least in developed countries. There are many reasons why our levels of physical activity have changed. For example, these days most of us work in office jobs which do not require hard physical labour. Similarly, the greater use of cars and car-driving has reduced how much we move.

But let's start at the beginning, how is physical activity defined?
Physical activity is defined as any movements and activity requiring our muscles, which we do as part of daily life. Most people think that physical activity is equal to exercise but that's not the case. Exercise is only one part of physical activity. Physical activity encompasses exercise but also includes all of our other daily movements. By contrast, exercise is more narrowly defined as planned, structured, repetitive and intentional movement, such as going for a run, yoga, exercise classes, swimming, etc. Some people completely abhor exercise and want to avoid it as much as possible. However, we can be physically active without doing exercise. We should therefore focus on being as physically active as possible in our everyday life. Of course, be my guest if you want to top this up further with an exercise session at the gym or swimming pool!

How much physical activity should we include in our daily life?
The public health recommendation is that adults should do at least 150 minutes per week of 'moderate physical activity'. Physical

activity is moderate if we can talk but not sing or whistle during the activity. At the moment, only 67% of men and 55% of women actually manage to do 150 minutes of physical activity. These rates further plummet as we get older. It's somewhat ironic that we are less physically active when we need it most – when we age.

One of the most common excuses for not doing enough physical activity is that we do not have enough time. Here then are some ways we achieve this goal more easily:

- **A little every day:** we don't have to be physically active for long extended periods to achieve our goal. It is actually better to have little bouts of 10 minutes' moderate activity dotted throughout our day or week. For example, instead of going for a long walk once a week, it is better to go for short 10-minute walks each day. Just get up and have a quick walk. Such short bursts can be incorporated into a hectic schedule. But even if we can't do those 10 minutes, simply minimise the amount of sitting down, which has a similar effect. Just being on our feet keeps us moving.

- **Start slow:** try not to overdo physical activity or exercise if possible. We have all been there: we have a New Year's resolution and are raring to go but often this resolution falters after a short period, or worse, we get injured. In sports science, the common saying is that one should not increase physical activity or exercise by more than 10% each time we increase our physical activity. So, if we start from very little physical activity, we have to take it very slowly to avoid getting injured. Injury means we end up spending an even longer period sitting around, which has a detrimental effect on our mental and physical health. Therefore, let's take it slow when we start being more physically active.

- **Strengthen yourself:** most of us have the impression that being physically active means doing cardiovascular-related activities, such as running, cycling, swimming, raking leaves or vacuuming the house. Cardiovascular activities are great but, ideally, they are complemented by physical strengthening activities. Physical strengthening means muscle and ligament strengthening, such as exercises with weights, yoga, pilates or even carrying shopping bags. Strengthening activities are important in complementing physical activities that improve cardiovascular health. Strengthening activity is important to avoid sarcopenia, a condition common in ageing where muscle mass is lost. Scientific evidence shows that physically inactive people lose around 3-5% of their muscle mass each year after the age of 30 with muscle loss really accelerating in older people. Why is muscle loss such an important issue? If we experience severe muscle loss as we age, we increase our risk of dying by three-fold. Scientists are still investigating the reasons why mortality increases so markedly, but we know that less muscle mass can lead to falls, which can in turn lead to other health conditions. Doing some strengthening physical activity twice a week could cut our risk of dying earlier by more than two-thirds. Strengthening physical activity is therefore definitely worth exploring.

- **Intensity vs consistency:** last but not least, we often think that physical activity or exercise must be intense to gain the best results. Intensity exercise has gained hugely in popularity in recent years, such as HIIT (High Intensity Interval Training) exercises, with good scientific evidence that it improves our physical health. However, working with great intensity can increase the chance of getting injured. We do not need to do physical activity or exercise at high intensity to stay healthy as we age. Instead, it is the amount of physical activity we do which is the critical factor. Consistency trumps intensity.

For example, many people have heard of the recommendation that we should do at least 10,000 steps a day. While there is little scientific evidence as to why this is the magic number for physical activity, there is excellent scientific evidence that if we do more steps, we reduce our risk for future disease. How many steps we need to do depends a lot on our physical health and what level of physical activity we start from. Recent studies have shown that the recommended number of steps might need to be more flexible, depending on our age and gender. For older people, the beneficial effects of steps seem to flatten or plateau when we hit 7,500 daily steps. Pushing ourselves to 10,000 steps might not provide additional benefits but potentially increase our risk of injury. The other key finding from these studies is that the intensity, how fast we walk, does not matter for gaining the health benefit. In other words, we do not need to walk fast to stay healthy, we just need to keep walking.

By following these tips, we can not only increase our healthy ageing lifespan we can also, incredibly, reduce our risk for Alzheimer's disease by up to 30% – just by being physically active.

What are the biological factors explaining why physical activity reduces our risk for Alzheimer's disease?
This is an area of a lot of ongoing research, but there are three key aspects to understanding how physical activity affects our risk of Alzheimer's disease:

1. Increased physical activity has been shown to reduce the accumulation of beta-amyloid by reducing the activity of the gamma-secretase, as well as increasing the clearing of beta-amyloid. Physical activity also reduces tau phosphorylation.

2. Increased physical activity reduces inflammatory processes in the brain. Inflammation can 'fan the fire' for beta-amyloid and phosphorylated tau production and accumulation. Hence,

reducing inflammation by exercising also slows down the processes associated with Alzheimer's disease.

3. Physical activity has been shown to reduce oxidative stress in the brain. Oxidative stress is a physiological process caused by molecules called free radicals. Free radicals can either be produced by the body itself or come from the environment, such as air pollution. The free radicals increase the oxidative stress of the cells which, similar to inflammation, 'fans the fire' for amyloid and tau production. Exercise has been shown to reduce the oxidative stress and therefore the milieu for Alzheimer's disease.

One final, more indirect, impact that physical activity has on the development of Alzheimer's disease is that it influences insulin regulation. We know already that faulty regulation of insulin, such as in Type II diabetes can increase our risk for Alzheimer's disease. Taken together, the impact of physical activity is enormous on our physical and mental well-being. Not only does it significantly reduce our risk for Alzheimer's disease but also many other diseases such as Type II diabetes and hypertension. The message for healthy ageing is therefore clear: get up and get active, it is worth it.

Diet and nutrition

We have already mentioned in relation to the two previous risk factors how our diet and nutrition might increase our risk for Alzheimer's disease. 'You are what you eat' is a proverb that makes it clear that what we put in our body will affect not just our physical but also our mental health. But it's not just important what we eat but also how much we eat.

We already talked about the recommended level of physical activity, but what is the recommended intake for calories?
This seems like a very straightforward question, but the correct answer depends on your body height, body weight, levels of physical activity, as well as other environmental factors, such as heat, cold,

etc. Still, on average, public health guidelines recommend that women consume 2,000 calories per day, and that men consume 2,500 calories per day.

Despite this sounding a lot, it is fairly easy to consume this with high-calorific snacks. For example, having a muffin with our coffee can set us back by 350-400 calories, up to 20% of our recommended daily calorie intake. Once we realise how calorific most of our everyday food is, it should become clear that it is quite easy to overshoot the recommended daily calorie intake, without even noticing. This might not be a problem if it happens occasionally or if we are very physically active but for most of us, these additional calories add up over time and fill out our waistline.

Before we explore how diet and nutrition affect our risk of Alzheimer's disease, we need to understand two other key issues for how our body's metabolism works. The first is that most of the calories we consume are used for our resting metabolic activity – that is, keeping our body functioning normally. The normal functioning of our body requires a lot of energy. For an average adult, the resting metabolic activity will be around 70-90% of our energy expenditure. Of that, 20% is used by the brain alone which is one of the highest energy consumers of the body.

Besides the resting metabolic activity, the remaining 10-30% will be used for physical activity, in particular muscle activity. This means that actually only a fairly small percentage of our daily energy consumption is used for physical activity. This often comes as a surprise to people, as they think that physical activity must require far more energy and that doing more physical activity or exercise allows us to consume more calories. In other words, we can 'burn off' those calories if we move enough. But now we know that it is very hard to 'burn off' calories simply by doing more physical activity or exercise, as it will only affect a small percentage of our daily metabolic activity. We have to be highly physically active to shift this balance which is not feasible for most of us. Instead it is far more effective to reduce our calorie intake.

The other key aspect that we need to understand is that our metabolic rates change with ageing. People often become aware that when they hit their 30s their weight seems to increase for no reason. They eat and are as physically active as before but somehow the weight seems to pile on more easily. I definitely experienced that myself, as I put on a few pounds/kilograms during that time, without even noticing that I got a bit 'soft around the edges'. For most of us, this increase in weight continues during our 40s and 50s with bulging waistlines often putting us in the overweight or even obese weight range.

What has happened? We seem to be eating and moving a similar amount as before but now we have all this additional weight, how can this be explained?

The reasons behind this common weight gain over your age are manifold. For example, our work-life might make us more stressed. Stress produces a hormone called cortisol in our body, which encourages the body to store fat, because in the past a stressful situation usually presaged a shortage of food. Today that is rarely the case but many of us will still experience chronic stress which tends to increase weight gain.

Even if we are not more stressed, there is consistent evidence that we are less physically active the older we get. Most of us hit our peak of physical activity metabolism in our late teens and early 20s. Our physical activity metabolism starts to reduce in the late 20s and gradually declines during ageing. We move less and our reduced physical metabolism reflects that. Hence, burning off calories will be much harder as we move less, even though we often think that we move as much as when we were younger.

Finally, on top of that, even our resting metabolism slows as we age. Let's remember that your resting metabolism accounts for 70-90% of our energy consumption. Scientific evidence shows that our resting metabolism reduces by 10% each decade after our 20s. These factors explain why we find it harder to shake off additional weight when we age. Still, it is not only how much we eat but it also matters 'what' we eat. Reducing our portion sizes is a good thing, as long as we stay within the healthy weight range, but even better would be to eat more healthily in general. This approach would give us additional benefits, both for our bodies and, more specifically, for our brains.

One could write a whole book about which food is beneficial for the brain and indeed many books out there do just that. But that's not the purpose of Tangled Up. Instead of going through all the food components which can benefit your brain, I will focus on the diet that has shown to reduce our risk for Alzheimer's disease – the Mediterranean diet. Many balanced diets are good for our brain health but over the last few years the so-called 'Mediterranean diet' has emerged as the most important one for the prevention of Alzheimer's disease.

What is the Mediterranean diet?
The Mediterranean diet refers to the traditional food consumed in Southern European countries. It's traditionally high in fruits, vegetables, particularly leafy ones, not starchy ones like potatoes, legumes and pulses such as lentils, beans. It also includes a moderate intake of oily fish such as sardines, white meat (such as chicken), as well as nuts and dairy. By contrast, there is a low intake of red meat (beef, pork), sugar and saturated fats – most fat in the diet comes in the unsaturated form of extra virgin olive oil, which is consumed in large quantities. Another critical factor for the Mediterranean diet is that – traditionally – alcohol tended to be consumed moderately with meals and not by itself, something which seems to diminish the negative effects of alcohol on our health.

If the Mediterranean diet reduces our risk for Alzheimer's disease, does that mean that Alzheimer's disease rates are lower in Mediterranean countries?
The surprising answer is no. The main reason is that although the 'traditional' Mediterranean diet is regarded as being good for our brain health, the modern Mediterranean diet is similar to all Western diets, very grain- and red meat-based. By contrast, the traditional Mediterranean diet was determined by what was readily available for people to consume – citrus fruits, olive oil, sardines, etc. But in modern Mediterranean societies there has been a shift away from this diet over the last few decades. This change in diet, along with a reduction in physical activity, was first noted in the 1960s in Mediterranean countries when the rates for cardiovascular disease

started to soar. Many people began to develop health conditions affecting their heart and blood vessels, such as atherosclerosis, which were previously not very common in those countries.

Public health organisations in Mediterranean countries realised this increase in cardiovascular disease in the population was probably due to the change of diet and a reduction in physical activity. Several research studies and clinical trials in Spain in the 1990s and 2000s, investigated whether transferring people to a more traditional Mediterranean diet might make a difference to their health. Their findings were astonishing, as the people who were told to follow a more traditional Mediterranean diet cut their risk for cardiovascular disease by 30%. To put this in context, such a large risk reduction is similar to that found in people on medication designed to lower cardiovascular risk, such as statins. Sticking to this diet might remove the need to take cardiovascular medication in the future. These astonishing findings were replicated by several subsequent studies, providing further evidence that a more traditional diet improved Mediterranean people's health outcomes.

How is this relevant to Alzheimer's disease?

We know already that cardiovascular health is a risk factor for Alzheimer's disease. For example, blood pressure hypertension or atherosclerosis increase our risk for Alzheimer's disease. It would, therefore, make sense to reduce this vulnerability by reducing our cardiovascular risk. Or to put it differently: 'What is good for the heart is good for the brain.'

The effects of a traditional Mediterranean diet on the development of cognitive symptoms and Alzheimer's disease was first explored in study in Finland, the FINGER trial. It might seem surprising that this study took place in Finland but its aim was to explore whether a Mediterranean diet could be adapted to a non-Mediterranean country with a completely different diet. The results of the FINGER trials showed reduced cardiovascular risk in its Finnish participants. Unsurprisingly, there are now many trials in the world taking the approach of these early studies and changing people's diets to be more Mediterranean-like.

The results so far are extremely promising and contrast starkly with so many negative findings for the medication studies. Not surprisingly, there has been a shift in recent years towards recommending people eat well and move more, by changing their lifestyle instead of waiting for medication. But before you throw your pills in the bin, switch to a Mediterranean diet and be more physically active, please consult your doctor.

A final word on some 'alternative food treatments' for dementia, which have been circulating for a number of years now. Most of these alternative treatments which claim to 'cure' or prevent dementia have little or no scientific evidence to back them up. To quote Mark Twain: 'A lie can travel halfway around the world, while the truth is putting on its shoes.' To stay on the safe side, we need to be cautious when hearing of 'miracle cures'. If you are unsure whether this is something you or your loved ones should try out, please talk first to your doctor or consult the dementia charities in your country, who can provide reliable information.

Finally, let's explore the fourth modifiable risk factor for Alzheimer's disease – sleep.

Sleep

Sleep is an incredibly important part of our lives without us often realising it. We spend around a quarter of our lives sleeping (200,000-250,000 hours), so there must be a good reason for us to spend such a long time not being active. Indeed there is, our whole body uses our sleep time to restore or 'tidy up' the mess we made during our wake time. Like a disgruntled parent tidying up after their beloved children once they have gone to bed (I might be projecting here), sleep time means tidying up before the 'madness' of daytime starts all over again.

What does the tidying-up process mean for the body?

Tidying up means restoring our cell functions, removing cell waste products and re-establishing our general physiological functions. Sleep is therefore vital to keep us healthy and there is a large body of scientific evidence showing that if our sleep is disrupted for a prolonged period, it has significant impacts on our overall health, including our brain health.

Before we dive into how important sleep is as a risk factor for Alzheimer's disease, we need to understand a little bit more about how sleep works in the brain. Until the 1950s we thought that during sleep our brain simply shut down, and restarted in the morning when we woke up. Now we know that instead of our brain 'shutting down', the brain has varying levels of activity during different stages of sleep. These stages form a specific series, with each complete series forming a sleep cycle. During a normal night, we go through five sleep cycles, each lasting around 90 to 110 minutes, which adds up to around 7.5 – 9 hours of sleep. In each sleep cycle there are five sleep stages, which are divided into non-REM sleep and REM sleep (REM stands for 'Rapid Eye Movement'). Stages 1-4 are non-REM sleep and stage 5 is REM sleep.

Sleep stage 1 starts right off within minutes, if not seconds, when we nod off and is also referred to as a light sleep stage. Light sleep means we are still somewhat alert and can be easily woken without feeling too drowsy. 'Catnaps' during the day are often stage 1 sleep, which a lot of people find very refreshing and easy to wake up from. The reason for this 'refreshing' feeling is that our brain slows down by switching its activity to very slow alpha and theta brain waves and our eye movements slow down at the same time. At the same time, our whole body and muscles relax as well, with the occasional muscle jerk which we often experience when drifting off to sleep. Stage 1 is quite brief and lasts only a few minutes before we enter the next sleep stage.

Sleep stage 2 is still fairly light, which means we can still be woken up easily. However, unlike Stage 1, we have sleep spindles and K complexes in our brain activity during Stage 2 sleep. Sleep spindles are nothing more than brief bursts of neural activity. K complexes are similar bursts of activity but much larger than sleep spindles. A critical function of sleep spindles and K complexes is that they help us to lay down our memories, in a process called memory consolidation (from Latin consolidate = 'to make solid together'). Sleep is critical for memory consolidation with particularly sleep spindles in stage 2 having an important role in this function. We also know that there is a reduced number of sleep spindles in people who have Alzheimer's disease when they fall asleep but it is currently not known how these changes to our sleep spindle bursts affect our memory consolidation in Alzheimer's disease. The other function of sleep spindles and K complexes is that they facilitate the process of 'drifting off' into deeper sleep, as they are both important for suppressing brain activity and arousal. Basically, sleep spindles and K complexes offer the brain protection by deterring it from waking up and they do this by suppressing activity in the brain. During sleep stage 2, body temperature falls and our heart beats slower. We are now ready to enter the deeper phases of sleep.

Stages 3 and 4 are usually combined into one stage, since they can look very similar. Now, we have entered deep sleep and our brains produce much slower brain activity, called delta waves. This brain activity is like the gentle swell of ocean waves. In sleep stages 3 and 4, it's much harder to wake us up. The brain is now disconnected from the outside world and the body can get started on its restorative work. For example, muscle tissue gets repaired, the immune system gets a boost and, importantly for Alzheimer's disease, nerve cells are repaired and cleaned of waste products. Basically, a big tidy up has started to prepare our body for the next wake time. After tidying up we can enter the world of dreams.

In sleep stage 5 – the REM (Rapid Eye Movement) sleep – we enter our world of dreams. Sleep stage 5 starts after we have been through all the previous sleep stages (1-4). Sleep stage 5/REM sleep can last up to one hour and is the longest sleep stage. Our brain becomes more active once more and the nerve cells start firing again as their processing work speeds up. This increased brain activity produces our dreams. While nearly all of us dream, the functions of dreams are still not understood. But physiologically we know very well what happens in REM sleep. During this stage, our body is completely motionless – in fact, our brain activity suppresses movement in our muscles. The reason it does this is to protect us from 'acting out our dreams', which can be dangerous. Better therefore to have our muscles relaxed and just have all the action happening in our brain. Nevertheless, sometimes the suppression of muscle activity doesn't work perfectly and we might still act out our dream, scaring the wits out of anyone nearby. It is completely normal to have these occasional 'acting out dreams' episodes, but when they become more common it can actually be a sign of another type of dementia – Dementia with Lewy body. But it is usually not an issue for Alzheimer's disease.

We now know that our brain suppresses muscle activity during REM sleep, except for one muscle group – those found around our eyes. The term REM – Rapid Eye Movement – refers to the fact that while our eyelids remain closed during this sleep stage, our eyes themselves move in a rapid and jerky way in different directions. The relevance of REM sleep in Alzheimer's disease is still being explored. In common with sleep stage 2's spindles, REM sleep plays a role in learning and memory, but its specific function is not yet clear. More importantly, it is unclear whether this has any relevance for the loss of memory seen in Alzheimer's disease. After the REM sleep stage we often have a brief lull in our sleep before we return to stage 1 sleep. It is usually during this transition, from REM sleep to stage 1, that we might wake in the middle of the night before drifting off back to sleep again. When we age, we sleep less than when we were younger. While children and adolescents require nine to 10 hours of sleep per night, working-age adults need eight hours of sleep. Finally, in older age, six to seven hours of sleep is the average for most people, besides the occasional nap during the day. It is somewhat puzzling therefore that we need the least amount of sleep when we age, while our body actually needs the most 'restorative work' when we are older. The reasons for this are unclear but we need to reassure ourselves that it is completely normal to sleep less when we age. Other sleep changes we see during ageing, such as waking up early and having to go to the bathroom, are also completely normal, and should not worry us. Clearly, if we are concerned about our sleep changes we should talk to our doctor but for most people these sleep changes are simply part of the healthy ageing process.

More problematic are conditions which affect our sleep directly. For example, obstructive sleep apnea is a condition where a person pauses to breathe or has periods of shallow breathing which causes disrupted sleep. Obstructive sleep apnea is a common condition, in particular for men over 40 who are overweight or those who drink too much alcohol. Sleep apnea is caused by the soft tissue in your throat, such as the tongue and soft palate, temporarily relaxing and narrowing or even closing the airway – restricting breathing for a short period. Relaxing the muscle leads to loud snoring – to the annoyance of our bedroom partner – as well as waking up and gasping for air. People who have sleep apnea report therefore a very poor sleep quality and feel very tired during the day. More worryingly, long-term sleep apnea can lead to other health conditions, such as high blood pressure, higher chance of stroke or Alzheimer's disease. Again, if you have the occasional bad night or you're snoring, there is nothing to worry about but if this is a common occurrence you should see your doctor, as it can affect your sleep quality and increases your risk for disease, such as Alzheimer's disease.

What about shift work?
Shift working is quite common across many professions, However, shift working really disrupts our sleep cycles since we are sometimes awake when we should be asleep and asleep when we should be awake. This is especially the case if we are alternating between day, evening and night shifts. Large studies have shown that long-term shift workers, defined as over 20 years of shift work, have a significantly higher risk for Alzheimer's disease. Ironically, this also affects nurses, doctors and care personnel who often look after people with dementia. Most night shift workers sleep on average only five to six hours per whole day, which means that they are losing one to two whole sleep cycles per day. If we do shift work for only a few years that might not be a problem but if we lose up to two sleep cycles over decades the scientific evidence is clear that we will have a higher risk for Alzheimer's disease in the long-term.

Why does prolonged irregular or disrupted sleep influence our risk for Alzheimer's disease?
The exact reasons are still being explored. However, preliminary scientific evidence shows that irregular and disrupted sleep affects the clearance of beta-amyloid and phosphorylated tau. We already

know that the brain removes waste products during sleep stages 3 & 4. This waste product removal includes the removal of beta-amyloid and phosphorylate tau. Recent studies have shown that during sleep stages 3 & 4 beta-amyloid gets 'flushed out' of the space between the nerve cells where they accumulate. It would make sense then that a disruption or even just a reduction in our sleep stages 3 & 4 could cause increased amyloid accumulation as less amyloid is being 'flushed out' from the brain. But future scientific studies are needed to confirm this. Right now, all we can say is that long-term disrupted or irregular sleep leads to a higher risk of Alzheimer's disease.

Other modifiable lifestyle risk factors

In the above sections we explored the main risk factors for Alzheimer's disease. However, there are several others which we could not cover. For example, cigarette smoking, traumatic brain injury, air pollution, reduced social contact and depression have also been shown to increase our risk for Alzheimer's disease. Ultimately, there are so many potential risk factors which we could take into account but having too many things to worry about might do as much harm as good. I have therefore focused this chapter only on some major risk factors. If you can control those major risk factors we covered, the scientific evidence is unequivocal in that you will significantly reduce your risk for Alzheimer's disease.

Let's stop worrying and start changing our lifestyle: it will make a big difference to how we age.

PART 4.
SUMMARY

In this part of the book, we have covered the following aspects:

- There are many risk factors for Alzheimer's disease but while some are not changeable quite a lot of others are.
- The genetic risk for Alzheimer's disease is in general very low. The risk is particularly low for genes which will definitely cause Alzheimer's disease as those are caused by mutation within specific families.
- There are other risk genes for Alzheimer's disease. However, all of these will 'only' increase or decrease the risk for Alzheimer's disease.
- Genetic testing needs to be carefully considered – even for risk genes – as it might not only affect our healthcare cover and employment but also our family, including children.
- Lifestyle risk factors can be divided into modifiable (those we can change) and non-modifiable (those we cannot change).
- Non-modifiable risk factors for Alzheimer's disease include our age, level of formal education early in life and hearing loss during middle age. There is an ongoing discussion as to

whether formal education levels and hearing loss at middle age are modifiable to some degree.

- Modifiable lifestyle risk factors for Alzheimer's disease have become increasingly important as the development of new medication has lagged behind.
- The good news is that modifying our lifestyle can reduce our risk of future Alzheimer's disease by up to 40%
- The top five modifiable lifestyle risk factors are: diabetes, hypertension and obesity; physical activity; diet and nutrition; and sleep.
- Diabetes, hypertension and obesity should be considered together because they are closely related. Keeping a healthy weight will not only reduce our risk for diabetes Type II and hypertension but also Alzheimer's disease.
- Physical activity is another key modifiable risk factor and by itself can reduce our risk for Alzheimer's disease by up to 30%. Physical activity encompasses any everyday activities we are doing and not just 'exercise'. Being as physically active as possible, regardless of the intensity or duration is key for health ageing.
- Diet and nutrition are important for our cardiovascular health. The most promising diet for the prevention of Alzheimer's disease is the Mediterranean diet, which limits consumption of grain and red meat.
- The final lifestyle risk factor is sleep. Sleep allows our body to regenerate and 'tidy up' our cells, including our brain cells. Sleep phases have been shown to increase the removal of beta-amyloid, while disrupted sleep can reduce its removal. Having good, uninterrupted sleep is key for good brain health.
- Other risk factors, such as smoking, depression and air pollution, also increase our risk for Alzheimer's disease. However, if

we focus on the top four risk factors, we can reduce our risk for Alzheimer's disease by a total of 40%.

- The important take home message is that we can take control of many risk factors for Alzheimer's disease. Instead of waiting to see whether we will develop Alzheimer's disease, modifying those factors allows us to take active control. We will also live a healthier life.

PART 5.
RARER FORMS OF ALZHEIMER'S DISEASE

The previous parts of the book introduced us to many aspects of Alzheimer's disease, but so far, we have 'only' covered the most common form of Alzheimer's disease. This is the form of Alzheimer's disease that accounts for around 80-90% of all Alzheimer's disease, has the classic episodic memory and spatial orientation symptoms and is therefore relevant to most people when they talk about the disease. But despite the rarer forms being less well-known and researched, they have a similarly serious impact on the people with dementia and their families. Covering those rarer forms of Alzheimer's disease not only informs the wider public of these different presentations of Alzheimer's disease, but also healthcare professionals who might have less knowledge of these forms.

The label 'rarer' refers simply to the fact that on a population level these forms of Alzheimer's disease occur much less often. However, 'rarer' should not imply 'of less importance'. Quite the opposite. People with these 'rarer' forms of Alzheimer's disease are often the most in need of support as it often takes a long time for them to receive a diagnosis and even after diagnosis most dementia support is geared towards the most 'common' form of Alzheimer's.

Scientifically, rarer forms are important as they provide a different window on the underlying disease. In fact, contrasting the symptoms and disease progression in rarer forms of Alzheimer's disease with the common form not only helps to understand these rarer forms but also the most common variant. In particular, it gives an insight into how the disease might start in different ways in different people, something which has important implications for diagnosis and treatment for Alzheimer's disease as a whole.

One final point to clarify before we go through the rare forms of Alzheimer's disease is that all these different types have the same pathophysiological changes as the common form. In other words, all the rare forms have an accumulation of beta-amyloid and phosphorylated tau as key pathophysiological features. Since the rare forms have the same pathophysiological changes as the common form, they are considered Alzheimer's disease and not a different type of dementia.

What then is the difference between the common and the other rarer forms of Alzheimer's disease if they have the same beta-amyloid and phosphorylated tau changes?

The key difference is often where in the brain the amyloid and tau start accumulating. The region in which the proteins accumulate and cause nerve cell loss will determine the symptoms people develop for the disease. In essence, the underlying disease is the same, but it is the symptoms which differ between the rarer and the common form of Alzheimer's disease.

36.
FRONTAL VARIANT ALZHEIMER'S DISEASE

The first of the rare forms of Alzheimer's disease that we will explore is the frontal variant of Alzheimer disease, also sometimes referred to as dysexecutive Alzheimer's disease. For this rare form of the disease people usually have some degree of episodic memory problems but at the same time show significant frontal or dysexecutive symptoms.

What are frontal or dysexecutive symptoms?

To understand the terms frontal and dysexecutive we must look at the brain's anatomy. Let's remember that the most common form of Alzheimer's disease affects the medial temporal lobe in the brain, at the centre of which is the hippocampus. To briefly recap, we already know that the accumulation of amyloid and tau in the common form of Alzheimer's disease occurs in the medial temporal lobe and the hippocampus when the disease starts. Since the hippocampus and medial temporal lobe are so important for episodic memory and spatial navigation, memory and orientation problems are the first symptoms in the common form of Alzheimer's disease.

In the frontal variant of Alzheimer's disease, amyloid and tau accumulate more in an adjacent brain region called the prefrontal cortex. The prefrontal cortex – also referred to as the frontal lobe, hence the name frontal variant Alzheimer's disease – lies in front of the medial temporal lobe. Most of it is located behind our eyes and forehead. In contrast to the medial temporal lobe region, which is quite a small, the prefrontal cortex is a vast region of the brain with many different functions. Not only is the prefrontal cortex important for initiation and coordination of our movement and spoken language it is also crucial for behaviour and decision-making. For frontal variant Alzheimer's disease, the behaviour and decision-making functions of the prefrontal cortex are particularly badly affected.

The prefrontal cortex initiates but also inhibits our actions or decisions. It also organises our thoughts, actions and language so that they are done in the right sequence. Another part of the prefrontal cortex controls our empathy towards others and processes moral and ethical decisions, ie, doing the right thing at the right time. Finally, the prefrontal cortex is strongly involved in how we perceive rewards and punishments, which shape our future actions and decisions.

Since this large brain region has so many functions, where amyloid and tau accumulate in the prefrontal cortex will determine which symptoms present. This is an important point to grasp and also explains why frontal variant Alzheimer's disease can be difficult to recognise and diagnose. People with frontal variant Alzheimer's disease can have a variety of symptoms. Most common are changes to the executive system of the prefrontal cortex – organising our thoughts, language and actions. When people develop problems with their executive system they are referred to as having 'dysexecutive' ('dys-' from Greek = abnormal, imperfect) symptoms.

People with frontal Alzheimer's disease often present at a doctor with disorganised thoughts, language or actions. Disorganised thoughts are by far the most common symptom, with people getting confused as to what they want to say or remember. As with the attentional memory problems we discussed in Part I of the book, the prefrontal cortex also helps to retrieve information from the hippocampus. So, if the prefrontal cortex is affected by Alzheimer's disease, people often also have memory problems but those problems have more impact on the laying down of those memories or their retrieval, than the actual indexing or storage of the memories. People having these kinds of dysexecutive symptoms often seem rather distracted and muddled, flitting from one subject to another.

Another common symptom in frontal variant Alzheimer's disease is a lack of motivation, so-called apathy (from Greek apatheia = freedom from wanting). Families often report that people are more likely to sit for prolonged periods, staring in front of them and no longer engaging in previously rewarding activities, such as hobbies. It is still possible to prompt people with apathy to do things. Nevertheless it is heartbreaking for the families that the person with these sorts of symptoms is less engaged, even in the activities they previously loved doing.

Other behavioural changes in frontal variant Alzheimer's disease are less common and, if they present, are milder. For example, people might become more disinhibited in their behaviour and language, doing or saying things out loud which they would have never done before. This often leads to embarrassment, so it is important to understand that the person is not purposely making inappropriate comments or actions, but rather the blame rests with Alzheimer's disease.

Finally, prefrontal cortex changes can also affect the regulation of our emotions. People with frontal variant Alzheimer's disease can therefore be more irritable or even aggressive, 'flying off the handle' with little or no reason. This misregulation of emotional responses can also go the other way in frontal variant Alzheimer's disease by making people more teary or emotional – for example, crying for no particular reason. Again, this can be very distressing for the person with Alzheimer's disease and their family, but we should remember it is more likely that these sudden emotional outbursts can be attributed to the prefrontal cortex changes, rather than the person in question actually being upset.

As we've already seen, these symptoms of frontal variant Alzheimer's disease often go hand in hand with episodic memory problems, but can even overshadow the memory problems. The main reason for this is that people will notice changes in personality and behaviour more than changes to memory. The focus on behavioural changes makes it much harder to get a diagnosis for frontal variant Alzheimer's disease, as clinicians will first investigate whether these behavioural changes might be due to family events, medication or mental health changes.

Other rarer forms of dementia which are not Alzheimer's Disease (such as the behavioural variant of frontotemporal dementia) also share the same behavioural changes. Behavioural variant frontotemporal dementia is caused by proteins other than the amyloid and tau which are involved in Alzheimer's disease. This means that the treatment approaches are quite different. Investigations are now underway to establish whether we can distinguish people with frontal variant Alzheimer's disease from other dementias by using Alzheimer biomarkers. Identifying Alzheimer's disease would then be less reliant on identifying the actual symptoms people present with and concentrate instead on which protein accumulations they are showing. Ultimately, we will be treating the protein accumulations which should stop or even resolve the symptom progression.

That's all very well for future treatments, but it's of little use now incaring for frontal symptoms in people who have behavioural and decision-making changes. There is considerable evidence that carer burden and distress is highest in these cases. The reason is that changes to behaviour and personality are not only much harder to accept, but it can be heartbreaking to see the person we love completely change their behaviour in front of our eyes. The behavioural changes also require all our energy to channel or 'manage' those behaviour and personality changes, with the person often behaving erratically, putting carers and families under more stress.

A piece of advice we often give to people who care for someone with frontal variant Alzheimer's disease or have such frontal symptoms is to consult the care advice for behavioural variant frontotemporal dementia (which, as we've mentioned, is not Alzheimer's disease). Although frontal variant Alzheimer's disease and behavioural variant frontotemporal dementia are different in many respects, the behavioural and personality changes can be very similar. Therefore, the care and management techniques that we use in behavioural variant frontotemporal dementia may be more applicable in the care of frontal variant Alzheimer's disease, rather than the care advice used for the common form of Alzheimer's.

Going forward, there is a clear need for better diagnosis and care guidelines for frontal variant Alzheimer's disease. The new biomarkers for Alzheimer's disease should make a difference, as they will allow the identification of the underlying disease regardless of the presenting symptoms. As it stands, however, there is a dearth of information when it comes to caring for those with frontal variant Alzheimer's disease. This needs to be addressed, not only to alleviate the symptoms in the person with the disease but to ease the burden and stress on their families and carers.

We will see that this is a common problem for the rarer forms of Alzheimer's disease since much less is known about them. It highlights that more tailored advice for those groups is urgently needed to complement the burgeoning amount of information available about the common form of Alzheimer's disease. Let's explore the next rare form of Alzheimer's disease – posterior cortical atrophy.

37.
POSTERIOR CORTICAL ATROPHY

Posterior cortical atrophy (PCA) does not even have Alzheimer's disease in its name, but is also caused by the accumulation of amyloid and tau.

What does posterior cortical atrophy mean exactly?

First, posterior means 'at the back'. Cortical refers to the cortex – the more superficial areas of the brain. The term atrophy is the shrinkage of the brain through loss of nerve cells. In short then, posterior cortical atrophy affects superficial areas at the back of the brain – the so-called parietal cortex and occipital cortex in the brain.

In posterior cortical atrophy, those regions at the back of the brain are first affected by the disease and not the medial temporal lobe, as in the common form of Alzheimer's disease. Unlike the common form of Alzheimer's, people with posterior cortical atrophy consequently often have little or no episodic memory problems at the beginning of the disease. This lack of memory problems meant it took until the 1980s for people with the disorder to be recognised as having Alzheimer's disease. Even today many non-specialist are unaware of this rare form.

If memory problems are not evident in posterior cortical atrophy, what symptoms do they usually have?

To understand this, we must again explore the brain anatomy and function of the 'back of the brain'. The back of the brain consists of two large main structures, the occipital cortex and the parietal cortex, which lie next to each other. The occipital cortex is the part of our brain lying right at the back of our skull. The parietal cortex lies slightly above and in front of it. Both brain regions are critical for our visual and bodily senses.

First off, the occipital cortex processes all the input from our eyes and hence is important for us to 'see' the world around us. It might come as a surprise that this region doesn't actually sit close to the eyes at the front of the brain, but at the opposite end, as far away as possible from the eyes. It's slightly counterintuitive, but we can blame evolution for its construction. You can really notice how important the occipital cortex is for seeing when you bang the back of your head on a hard surface. Directly after hitting the back of the brain, most people report 'seeing' stars or other shapes. Clearly, we are not seeing those stars or shapes with our eyes. Rather, the impact on the occipital cortex has made it misfire, generating the visual image of these stars or shapes.

The occipital cortex not only 'sees' what the eyes see but also 're-assembles' the input from the eyes. Let me explain. Our eyes are not particularly clever: they only record what we see but cannot recognise or interpret what we see. That job falls to the occipital and parietal cortices. The occipital cortex puts together all the information received from the eyes and assembles a picture in our brain of the world we see. The parietal cortex takes the information from the occipital cortex and processes it further. For example, the parietal cortex links the visual information from the occipital cortex with our other senses, such as touch and hearing, as well as the areas of the brain associated with movement. The parietal cortex is therefore a key 'integration' brain region, literally integrating information from adjacent brain areas.

In posterior cortical atrophy, these processes of visual 're-assembly' and integration in the occipital and parietal cortices are disrupted or damaged. This can result in people developing problems with their vision – despite their eyes working perfectly well. Faulty visual functions in our brain, despite having intact eyes, can lead to the typical symptoms of posterior cortical atrophy such as 'cortical blindness' and 'visual neglect'. Cortical blindness means that we can develop some degree of blindness because of the changes in our occipital cortex, even while our eyes are working correctly. Now, blindness does not necessarily mean complete blindness. Instead, in posterior cortical atrophy, the cortical blindness presents more as 'blind spots' in our vision. In those blind spots we can no longer see certain parts of objects or scenes.

Visual neglect is the other major symptom caused by disruption or damage to the parietal cortex. Instead of developing blind spots for specific parts of visual scenes, we lack visual attention. This means that even if our eyes and occipital cortex are working correctly and we can 'see' whole scenes, our brain does not pay attention to certain parts of the scene. The most common form of visual neglect is that people can't see the right or left side of their visual field.

This is all a bit abstract, so let's have a look at an example. Nearly everyone knows Leonardo Da Vinci's painting of the Mona Lisa. But if not, have a quick look at the picture online. One of the most mysterious and fascinating things about the Mona Lisa is that the right and left sides of the picture are quite different, including the Mona Lisa herself. The Mona Lisa is famous for her mysterious smile and people argue over whether she smiles or not. Da Vinci deliberately tries to trick our parietal cortex by painting the right side of the Mona Lisa's face showing her smiling, while the left side of face shows her not smiling. Since we are switching our attention from the left to the right side of her face and vice versa, our brain gets conflicting inputs that sometimes the Mona Lisa seems to smile while sometimes she does not. Leonardo Da Vinci brilliantly tricked our brain to keep us endlessly fascinated by the picture. The picture also provides a great way of understanding visual neglect, since it so deliberately splits the left and right side of the picture in half.

For someone with visual neglect, the Mona Lisa will be always smiling or not smiling when they look straight at the picture. Whether they see her smiling or not depends on whether the left or right parietal cortex is more affected by the disease. The key point of this example is that a person with visual neglect cannot shift their attention from one side of their visual field to the other and therefore 'neglects' one particular side. In contrast, someone with partial cortical blindness would still see the Mona Lisa alternating between smiling and not smiling. They would, however, have problems seeing complete parts of the painting. For example, they would not see no longer see the Mona Lisa's hands or even her face. There would be simply a black spot in that location.

Both cortical blindness and visual neglect are key symptoms of posterior cortical atrophy. As we've already seen, the accumulation of amyloid and tau causes posterior cortical atrophy. Likewise, the accumulation of amyloid and tau in the occipital and parietal cortices causes the visual symptoms in posterior cortical atrophy. Since the accumulation of amyloid and tau occurs slowly, the visual symptoms in posterior cortical atrophy emerge slowly and are subtle at first. These subtle visual problems at the beginning of the disease make people with posterior cortical atrophy think that something must be wrong with their eyes. It is common for people in the earliest stages of posterior cortical atrophy to go to their optician to check whether they need new glasses or see an ophthalmologist to check if they have other problems associated with ageing, such as a cataract. Of course, we now know that a new set of glasses or a cataract operation will not help for posterior cortical atrophy, since the symptoms are not caused by the eyes but by how the occipital and parietal cortices process the input from their eyes.

How can we spot the symptoms for posterior cortical atrophy?

Problems with vision are very common during ageing, so we should not become too concerned when we develop changes to our vision. Visiting our optician or doctor is a good first step to get our eyes checked and find out whether the visual changes originate in our eyes. If they do, then either a new set of glasses or some other medical procedures can likely rectify that problem. However, if this does not improve the symptoms then it might be worth discussing this with your family doctor to decide whether a referral to a neurologist or neuropsychologist might be warranted.

Neurologists or neuropsychologists have more systematic tests, similar to my Mona Lisa analogy, which can check whether we can see a whole picture or miss parts of one. They may also test your reading of written text, as people with posterior cortical atrophy find it increasingly hard to read texts. The reason for this is that reading requires tracking letters, words and rows down a page. Many people with posterior cortical atrophy quickly become confused as to which line they are on and the lines seem to become a mess. The reason for this is again our visual attention, which helps us to keep track of the information we read, as you are doing now.

Finally, a common symptom in posterior cortical atrophy is bumping into furniture and door frames for no apparent reason. Again, the visual problems found in people with posterior cortical atrophy means that they cannot see the objects they're colliding with. Luckily, such 'bumps' largely cause only bruises but on occasion they can lead to more serious injuries. Such knocks can also bruise mental health, as people with visual problems become very hesitant in their movements, since they are aware that they might hurt themselves.

A final aspect of posterior cortical atrophy we need to mention is that this rare form of Alzheimer's disease is considered a 'young-onset dementia'. Young-onset dementia means that people start their first symptoms before the age of 65.

Why 65, you might ask?

The honest answer is that this is an arbitrary threshold. Nevertheless, it allows clinicians and scientists to distinguish between young-onset (< 65) and late-onset (> 65) dementia, as the vast majority of people with Alzheimer's disease will get dementia after the age of 65. The majority of people with posterior cortical atrophy display their first symptoms before the age of 65. It is not uncommon to see someone in their 50s with posterior cortical atrophy but it is less common to see someone older in their 70s or 80s developing this rare form of Alzheimer's disease.

In the next chapter, we will explore another rare form of Alzheimer's disease which is considered a young-onset dementia – logopenic variant primary progressive aphasia.

38.
LOGOPENIC VARIANT PRIMARY PROGRESSIVE APHASIA

Logopenic variant primary progressive aphasia is quite a mouthful, so let's start by explaining the terminology for this rare form of Alzheimer's disease. Logopenic comes from ancient Greek, combining the words logos = 'words' and penia = 'lack of'. So, logopenic literally means a 'lack of words'. Primary means that this lack of words is the first or main symptom in people with this variant of Alzheimer's disease. Finally, progressive aphasia (from Greek aphatos = 'speechless') is a syndrome describing worsening speech problems. Logopenic aphasia therefore literally means a 'speechless lack of words'. Many people come across the term aphasia when someone has a stroke affecting the speech areas of the brain and they have difficulties speaking. The difference between aphasia caused by stroke or Alzheimer's disease is that in stroke, the aphasia is a one-time event. This means the speech in the person is affected at once but the symptoms stay the same or improve after a while. By contrast, in progressive aphasia such as logopenic primary progressive aphasia, the speech and language symptoms slowly emerge and worsen over time. As in the case of posterior cortical atrophy, there is no mention of Alzheimer's disease in the term logopenic variant primary progressive aphasia. The reason for this is that for a long time it was not considered part of Alzheimer's disease at all. Instead, people with these symptoms were considered to have a non-Alzheimer form of dementia – frontotemporal dementia. That is because logopenic variant primary progressive aphasia features language problems very similar to – sorry about this – nonfluent variant primary progressive aphasia, which is a type of frontotemporal dementia. Even the experts have sometime difficulty distinguishing these subtypes.

In the early 2000s it was 'discovered' that logopenic primary progressive aphasia was actually Alzheimer's disease. The realisation emerged once biomarkers showed that these people showed the protein signatures of Alzheimer's. In other words, logopenic variant primary progressive aphasia is caused by the same proteins as all other forms of Alzheimer's disease – amyloid and tau.

People with logopenic variant primary progressive aphasia rarely have memory or spatial disorientation symptoms, as in the common form of Alzheimer's disease. What, then, are the specific language symptoms of logopenic progressive aphasia? In practice, the language symptoms are quite subtle, with most people having this form of Alzheimer's disease reporting difficulties with 'finding the words'. Doesn't this sound similar to the common form of Alzheimer's disease? Remember how in Part I of the book Alois Alzheimer probed the word-finding difficulties in Mrs Deter? However, when people with logopenic primary progressive aphasia report similar difficulties, it does not mean that they can't remember the words (indeed, their memory is usually intact in the early stages of the disease). Rather, what they describe is that they can no longer say the word out aloud, despite remembering it and knowing what it means.

The fact that people with logopenic primary progressive aphasia can remember the meaning of the words also allows us to distinguish them from those affected by other primary progressive aphasias which are not Alzheimer's disease, such as semantic variant primary progressive aphasia (also called semantic dementia), which is a form of frontotemporal dementia. We already know that people with logopenic variant primary progressive aphasia have problems saying words out loud while still knowing the meaning of the words.

By contrast, semantic variant primary progressive aphasia people can usually say the word aloud but no longer know the meaning of the word.

To test this difference, clinicians often ask such patients to repeat complex, multisyllabic words before asking what they mean. A person with logopenic variant progressive aphasia might have difficulty saying a multisyllabic word, such as 'stethoscope' or 'caterpillar', but they know what the words mean. The opposite is the case for someone with semantic variant primary progressive aphasia, who would be able to repeat 'stethoscope' or 'caterpillar', but would not know what either was. Another test is to ask people to repeat a complex sentence or follow a multi-step command. Ideally, the sentence or command should have a main and a subordinate clause, which puts more demand on our language functions. For example, a person is asked by their doctor to say the sentence, 'Pick up the paper after touching the table.' Repeating this quite innocuous sentence poses quite significant difficulties for people with logopenic variant primary aphasia. More importantly, they would struggle to perform the actions in the correct order.

This struggle to say a sentence and to perform the action in it correctly often comes as a surprise to family members. After all, with the paper lying in front of someone, it seems a very straightforward command to 'pick up the paper after touching the table'. However, for a person with logopenic variant primary aphasia, the sentence structure is very confusing as they need to keep in their mind that they need to touch the table first before picking up the paper. Most of them would pick up the paper first, while others would touch the table with one hand and with the other hand pick up the paper at the same time. Often, after executing this action incorrectly, they would even ask, 'Which way around was it again?' It is important to realise that they did not simply forget the instructions but had difficulties understanding the actual sentence.

What is happening in the brain when such symptoms occur?

We already know that logopenic variant primary progressive aphasia is caused by the proteins amyloid and tau which cause Alzheimer's disease. But we also know that where the amyloid and tau accumulate in the brain is critical when understanding symptoms. For logopenic variant primary progressive aphasia, amyloid and tau accumulate in a brain region called the temporo-parietal junction. This is a junction between two brain regions: the temporal cortex and the parietal cortex.

We have come across these regions previously in the book. The temporal cortex is a large brain region which includes the medial temporal lobe where the common form of Alzheimer's disease starts. In logopenic variant primary progressive aphasia, amyloid and tau accumulate where the temporal lobe meets the parietal cortex. We can remember when discussing how our sight works that the parietal cortex is important for the integration of our senses, including vision, sensation and hearing. However, for logopenic variant primary progressive aphasia it is not sight but the hearing part of language processing that is affected by the disease.

The temporo-parietal junction has a critical function for our spoken language as it is part of the so-called phonological loop.

What is the phonological loop?
When we read we can sometimes hear ourselves reading a sentence with our 'inner voice'. We do not say the words out aloud but we can 'hear them' inside our brain very clearly. This inner ear and voice is our phonological loop working. The phonological loop is necessary to keep track of what is being said or read. If we did not have this language short-term working memory, we would be unable to follow a conversation or a book as we would forget what we had just said or read. This linguistic short-term memory is very different from our episodic memory system which learns information about events or spaces. Because these are two very different memory functions, they are located in different areas of the brain. As we know, the medial temporal lobe is important for episodic memory. For the phonological loop it is the junction between the temporal and parietal cortices. Saying long words or understanding complex sentences requires an intact phonological loop.

Multisyllabic words, for instance, require us to keep different parts of a word in 'mind' before we say it out aloud. Of course, if we do not have any problems with our phonological loop we are not even aware of this internal rehearsal of information. However, as soon as

Alzheimer's disease affects the regions of the brain involved in the phonological loop, we will struggle with those words. Phonological loop deficits become even more apparent in complex or subordinate clause sentences, where we need to keep track of information, like 'who does what'. For example, the instruction 'picking up the paper, after touching the table', requires us to keep track of the fact that it is the table we need to touch first before picking up the piece of paper. This might be a trivial example, but it can have quite significant implications for safety instructions, such as 'Only disconnect the hose after closing the gas valve.'

Besides the phonological loop symptoms, people with logopenic variant primary progressive aphasia often mispronounce words slightly within sentences. Again, it is the phonological loop which is important for the 'assembly' of words in our sentences. Hence, if the phonological loop becomes faulty, we will struggle to string together sentences or even the syllables of words. Spotting such changes in language in everyday conversation can be very challenging, as we all mispronounce words from time to time or use the wrong word in the wrong context. So we should not be worried if this happens occasionally. However, if it becomes a more regular occurrence it would be advisable to see your doctor to discuss these symptoms.

Similar to posterior cortical atrophy, who often see a optician first, people with logopenic variant primary progressive aphasia often get referred first to a speech and language therapist. The speech and language therapist should notice that the language changes are due to deficits in the phonological loop.

To sum up, logopenic variant primary progressive aphasia should be regarded as a 'language type' of Alzheimer's disease, as its main symptoms are related to language problems rather than memory problems. We have now covered nearly all the rare forms of Alzheimer' disease, except for one – corticobasal syndrome.

39.
CORTICOBASAL SYNDROME

I have deliberately left corticobasal syndrome as the last rare form to explore. The main reason for this is that we know the least about this rare form of the disease but also that corticobasal syndrome can overlap with many other diseases. Indeed, strictly speaking, corticobasal syndrome is not Alzheimer's disease. However, corticobasal syndrome is caused by elements of Alzheimer pathophysiology, although those pathophysiological changes overlap with other neurodegenerative diseases. Welcome to the 'murky waters' of corticobasal degeneration.

What is corticobasal syndrome?

This seemingly innocuous question is actually quite difficult to answer. Corticobasal syndrome is – as the name suggests – a syndrome. A syndrome is a group of symptoms which often occur together (dementia is a syndrome). These symptoms may be caused by different diseases, which can make it difficult to determine which of them actually caused the symptoms. Confused? You are not alone. In essence, corticobasal syndrome can be caused by several different types of dementia, including Alzheimer's disease, frontotemporal dementia and progressive supranuclear palsy.

The corticobasal part of the name refers to the brain regions affected by this syndrome. The brain regions are the cortex (the 'cortico' part of corticobasal) and the basal ganglia (the 'basal' part of corticobasal), which lie in the deeper brain regions. We will return to the functions of these brain regions and how their impairment might contribute to the symptoms. But let's explore the symptoms of corticobasal degeneration first, since they define the syndrome.

What are the symptoms?
Most symptoms in corticobasal syndrome are related to problems with movement and may be similar to those found in another disorder of movement, Parkinson's disease. Indeed they are often referred to as 'Parkinsonian', meaning they are similar to symptoms in Parkinson's disease. Parkinsonian symptoms include a slowing of movement, called bradykinesia (from Greek brady = slow and kinesis = movement), stiffness of the muscles (clinically called rigidity) as well as a shaking of limbs (clinically called tremor). Beside these Parkinsonian ones, corticobasal syndrome symptoms are also more specific, such as limb apraxia or asymmetric limb rigidity.

Limb apraxia (from Greek a = not and praxis = doing) refers to the inability to perform a purposeful action or movement. Often such limb apraxia is more pronounced in one limb or the limbs on one side of the body. Limb apraxia can be quite subtle at the beginning and often reveals itself only when we are trying to do actions which require complex movements and coordination – for example, putting on our clothes. When dressing, people with corticobasal syndrome often struggle to get their arms in their sleeves and their legs into trouser legs. To most of us putting on clothes seems to require only very basic coordination. However, we forget that it actually requires a lot of coordination, not to mention fine-graded movements, to get our hands into the sleeves. People with limb apraxia find such actions extremely challenging, so difficulties putting on clothes can be a typical symptom in corticobasal syndrome. The brain region particularly responsible for the limb apraxia is the parietal cortex.

We have come across the parietal cortex already in the chapters on two rare forms of Alzheimer's, posterior cortical atrophy and logopenic variant primary progressive aphasia. We already know that the parietal cortex is important for our senses, such as visual attention (posterior cortical atrophy) or the phonological loop (logopenic variant primary progressive aphasia). However, another important function of the parietal cortex is to connect the regions of the brain that deal with movement and sensing. Connecting what we see with what we do is very important – just think about when we need to coordinate our movement in order to reach for a glass of water. Without our visual system telling our movement system how far we have to reach, it would be difficult to pick up the glass. Instead, we would probably overshoot and tip the glass over, or undershoot with our movement and grasp at an empty space. At the same time, the areas of our brain that pertain to movement constantly feedback to our visual and touch senses when we touch something. Both systems are in constant interaction about the location of our body parts.

The parietal brain region allows this seamless flow of information between movement and senses. We are not aware of these processes but they are critical for our everyday life. Limb apraxia is a classic example for when the parietal lobe is no longer working properly. It should come as no surprise, then, that people with limb apraxia have difficulties coordinating their movements as their feedback loop in the parietal cortex isn't functioning properly. No wonder it is suddenly hard to put on clothes.

The parietal lobe relays information from the senses to our motor system. Losing some of this sensory feedback can cause the other specific sign in corticobasal syndrome – limb rigidity. This is because if the feedback from the senses, such as touch, is not correct, the muscles may not relax appropriately and are somewhat stiff. Doctors often measure this stiffness by asking patients to relax their arms and then move the patient's arms passively. If we leave our arms as floppy as we can, the doctor should be able to move the arms around without any problems, however, when we have limb rigidity the doctor encounters some resistance from our muscles when moving the limbs.

Another important feature in corticobasal syndrome is that this stiffness often only affects one limb or the limbs on one side of the body. This can be explained by the fact that in the beginning stages of corticobasal syndrome, often only one side of the parietal cortex is affected. The sensory deficits in corticobasal syndrome can also sometimes manifest as sensory deficits for touch, so that the person has a reduced sense of feeling in their limbs, for example in their fingers or toes. The sensory deficits can vary greatly from person to person. Corticobasal syndrome also affects the basal ganglia, which lie deep in the brain. The basal ganglia are complex so we will stick to some key concepts.

The first key concept to understand is that the basal ganglia often act as relays or regulators of the interplay between our deep and superficial brain regions. If they are broken or defective it can cause all kinds of problems. Imagine, for example, the regulator for injecting petrol in your car being defective. Your car would still run but it would 'misfire' and fail to perform optimally. It is the same in the brain. Even if our superficial brain regions are working normally, changes in the basal ganglia can cause the whole brain to perform suboptimally.

The second key concept to understand about the basal ganglia is that, being relatively small, they connect to many superficial brain areas (a so-called 'one-to-many' relationship). So, a decrease in the basal ganglia can subtly affect many different brain regions. This can result in many different subtle cortical symptoms. However, it is actually the faulty basal ganglia which are causing the problems and symptoms. To understand this complex concept a bit better, let's make an analogy to a network of airports. Let's imagine that the basal ganglia are a major airport, like Heathrow or JFK, which have many flight connections to other airports. One day this major airport was badly disrupted – let's say, a power cut in one terminal. This power cut will affect all flights arriving and departing from that terminal to other airports. The power cut therefore has a knock-on effect on the other airports, affecting their functioning despite them having no disruption or power cut. Dysfunction of the basal ganglia can cause all kinds of different, knock-on symptoms. So people with corticobasal syndrome can present with a 'mosaic' of symptoms.

If you find this whole chapter on corticobasal syndrome confusing, do not worry, you are not alone. Even experts in the field struggle to understand the complexity of this syndrome and its symptoms. It is therefore no surprise that it often takes a long time for people with corticobasal syndrome to be diagnosed correctly. It is simply a very difficult disease to distinguish corticobasal syndrome from other diseases. The new Alzheimer's disease biomarkers may help people to get a quicker diagnosis, so that they can be treated accordingly. Just keep in mind that corticobasal syndrome can occur as part of Alzheimer's disease but mainly affects our movements.

40.
A FINAL WORD ON THE RARER FORMS OF ALZHEIMER'S DISEASE

Before we summarise the findings in this chapter, it is worth reiterating the fact that despite all the different symptoms and presentations of the rare forms of Alzheimer's disease, they are still caused by the accumulation of amyloid and tau. And, consequently, while the symptoms of these rare forms may be very different, they are all caused by Alzheimer's disease – with corticobasal syndrome being the exception as its origin is still being explored.

Despite the fact that all these rare forms are united under the heading of Alzheimer's disease, people with the rare forms of Alzheimer's disease and their families are, in my experience, often exasperated as it usually takes them much longer to be correctly diagnosed and treated. The main reason for this is that since these forms are so rare, only specialists are able to spot and diagnose them correctly, and it can take a long time for cases to reach specialists. The other source of frustration is that even after they do have a diagnosis, the support for these rare forms is less developed than for the common form of Alzheimer's diseases. Even at dementia support groups, which are highly supportive places, people with these rare forms and their families feel out of place, as no one else seems to be experiencing the same symptoms and challenges. If you are experiencing this, it is worth seeking out so-called 'rare dementia' or 'young onset dementia' support groups. These support groups not only include people with the rare forms of Alzheimer's disease but also people with other rare forms of dementia, such as Frontotemporal Dementia, Dementia with Lewy bodies, Progressive Supranuclear Palsy and others. Given that the symptoms in the rare forms of dementia often overlap, such groups can provide additional support to general dementia support groups.

Finally, the dearth of knowledge about the rare forms of Alzheimer's disease highlights how much more research is needed to understand them. Unfortunately, most research is still concentrated on the common form of Alzheimer's disease. Nonetheless, over the last few years there has been a growing recognition that people with these rare forms of Alzheimer's disease and their families deserve better information and greater support.

PART 5.
SUMMARY

In this part of the book, we have covered the following aspects:

- Despite the majority of people with Alzheimer's disease having the common symptoms of memory loss and disorientation, Alzheimer's disease has several forms.
- People with frontal variant Alzheimer's disease often show changes to their behaviour and decision making – for instance, by being more disinhibited or even reckless. Alternatively, they can be apathetic and show little motivation and initiative.
- Frontal variant Alzheimer's disease also impairs people's ability to sequence information in the right order and they often seem 'muddled' in their thoughts, language and action.
- Posterior cortical atrophy is a rare form of Alzheimer's disease where people develop problems that involve visual impairment, rather than the more common memory deficits.
- Posterior cortical atrophy can be regarded as a 'visual form' of Alzheimer's disease where people often have difficulties seeing things: resulting in them bumping into furniture or driving their car into things, for example.

- The other common symptom is visual neglect, which results in people only perceiving one part of their visual field.
- Logopenic variant primary progressive aphasia is the 'language' form of Alzheimer's disease. People with it often have problems saying multisyllabic words out loud or following complex commands. People still know what they want to say but have problems 'getting the words out right'. This is due to changes in the phonological loop.
- Strictly speaking, corticobasal syndrome is not Alzheimer's disease but its pathophysiology overlaps with Alzheimer's disease.
- People with corticobasal syndrome have difficulties performing and coordinating movements, such as dressing oneself.
- People with this form of dementia can show a 'mosaic' of symptoms. This is because the basal ganglia are responsible for connecting different parts of the brain.
- There is an urgent need to understand these rarer forms so that we can improve diagnosis and care.

ACKNOWLEDGMENTS

Well, there we are, you made it to the end of the book. I truly hope that it has given you a better understanding of the understanding of the history and science behind Alzheimer's disease.

By now, you will have hopefully understood why the typical memory and disorientation symptoms occur in Alzheimer's disease. How Alois Alzheimer and Auguste Deter (and the forgotten Oskar Fischer) contributed to the discovery of Alzheimer's disease. How amyloid and tau cause changes in the brain that lead to Alzheimer's disease and the current treatment approaches for the disease. How genetics and lifestyle choices influence our risk for developing Alzheimer's disease in the future. Finally, how the rarer forms of Alzheimer's disease have other symptoms depending on where the amyloid and tau accumulate.

If you know the answers to all these questions, congratulations, you are now officially an Alzheimer's disease science geek!

I hope you enjoyed the journey through the book, as much as I did and that the book has made you more knowledgeable about the disease. In the end, knowledge is power, so having this newfound power, I hope you will put your newfound knowledge to good use.

I would like to say thank you to so many people who made this book possible. First of all, my family, who had to endure my writing bouts on weekends, evenings and holidays, and kept me going. Then there were all the supportive people providing feedback on earlier versions – including lay people, carers of people with dementia, and people with early cognitive changes themselves. I am particularly indebted to Neil Foley, Joyce Hopwood, Willy Gilder and Cerilynne Higgins for their amazing feedback and work on the book. Finally, I could not have written the book without the previous literature available. There are too many scientific studies and books to mention but I want to highlight Prof Konrad & Ulrike Maurer who did a lot of work on Alois Alzheimer's life story. I relied heavily on their work. Similarly, I want to thank Prof Michel Goedert, whose publication on Oskar Fischer made me aware of this 'overlooked' person who was critical for the discovery of the disease. Thank you.

I would also like to thank my publisher Canbury Press (Martin Hickman, Gaby Monteiro) for their insightful comments on my drafts.